Springer Series in Materials Science

Volume 168

For further volumes:
http://www.springer.com/series/856

The Springer Series in Materials Science covers the complete spectrum of materials physics, including fundamental principles, physical properties, materials theory and design. Recognizing the increasing importance of materials science in future device technologies, the book titles in this series reflect the state-of-the-art in understanding and controlling the structure and properties of all important classes of materials.

Yoshinobu Aoyagi · Kotaro Kajikawa
Editors

Optical Properties of Advanced Materials

Springer

Editors
Yoshinobu Aoyagi
Global Innovation Research Organization
Ritsumeikan University
Shiga
Japan

Kotaro Kajikawa
Interdisciplinary Graduate School of Science and Engineering
Tokyo Institute of Technology
Yokohama
Japan

ISSN 0933-033X
ISBN 978-3-642-33526-6 ISBN 978-3-642-33527-3 (eBook)
DOI 10.1007/978-3-642-33527-3
Springer Heidelberg New York Dordrecht London

Library of Congress Control Number: 2013931747

Printed on acid-free paper

Springer is part of Springer Science+Business Media (www.springer.com)

Preface

This book is designed to explain fundamental optical properties of advanced materials, which are recently being paid much attention as topics of basic research and application. This book covers various optical and electrical materials: quantum structures of semiconductors, spintronics, photonic crystals, surface plasmons in metallic nanostructures, photonic metamaterials, organic materials, and magnet-optics. So far, there have been few books that summarize these materials and methods from the viewpoint of optical properties of advanced materials. These materials have their own peculiarities, which are very interesting in modern optical physics and also in applications, because the concepts appeared in the optical properties are quite different from those in conventional optical materials.

This book is designed for graduate students, researchers, and/or engineers who are interested in fundamental optical properties in such advanced materials. Chapter 1 deals with quantum structures. First, the basic theory of quantum structure is explained. After that, the fundamental optical properties of low-dimensional semiconductor materials are explained from analytical points of view, in addition to real application of those low-dimensional materials to devices. In Chap. 2, the basic theory and research topics on photonic crystals are summarized. Many new physical phenomena are discovered in the field of photonic crystals, in the last two decades. The photonic crystal materials will be more important in the field of optical advanced materials. The chapter focuses on the nonlinear optical properties of photonic crystals. Chapter 3 deals with surface plasmon photonics, which is a hot research field in nanophotonics. The surface plasmons give attractive optical properties that cannot be realized in dielectrics and semiconductors. It is used in wide range of research fields, such as optics, physics, chemistry, and biology. Chapter 4 describes optical metamaterials that show interesting optical properties originating from their higher order metallic nanostructures. We can obtain an optical medium with a wide range of refractive indexes: from negative to large positive. This allows us to realize special optical applications, such as super high-resolution microscopy and optical croaking. Chapter 5 reviews spintronics, which is a new type of electronics that uses the mutual control between magnetic and other physical signals such as electrical and optical signals. It is one of the most important research fields of advanced materials. Chapter 6 provides a review on liquid crystal optics. Liquid crystals are the most successful functional organic materials, and are widely used for the flat panel displays. The liquid crystals with chiral part spontaneously form periodic structures, which are useful for modern photonic applications. This chapter focuses on the photonic effects in the chiral liquid crystals. Chapter 7 describes a review on organic light emitting devices (OLED). OLEDs are promising optical devices for high-contrast display and illumination. This chapter mainly focuses on the OLED materials. In Chap. 8, the effect of magnetic field on optical properties, i.e., magneto-optical (MO) effects, is described. The dynamical MO properties are becoming more important, since the change in MO property is deeply relating with the dynamical behavior of spin in advanced materials, and therefore it plays the key role in the growing field of spintronics.

Authors will be very pleased if this book proves to be helpful for students, engineers, and scientists who would like to understand the fundamental optical properties of the advanced materials. Finally, we wish to express our appreciation to Dr. Claus E. Ascheron for his encouragement.

Shiga Yoshinobu Aoyagi
Yokohama Kotaro Kajikawa

Contents

Contributors

Yoshinobu Aoyagi Global Innovation Research Organization, Ritsumeikan University, 1-1-1 Noji-higashi, Shiga 525-8577, Japan, e-mail: aoyagi@fc.ritsumei.ac.jp

Katsuhiko Fujita Institute for Materials Chemistry and Engineering, Kyushu University, Kasuga, Fukuoka 816-8580, Japan, e-mail: katsuf@asem.kyushu-u.ac.jp

Shin-ichiro Inoue Advanced ICT Research Institute, National Institute of Information and Communications Technology (NICT), 588-2 Iwaoka, Nishi-ku, Kobe 651-2492, Japan, e-mail: s_inoue@nict.go.jp

Kotaro Kajikawa Interdisciplinary Graduate School of Science and Engineering, Tokyo Institute of Technology, 4259 Nagatsuta, Midori-ku, Yokohama, Japan, e-mail: kajikawa@ep.titech.ac.jp

Shin-ya Koshihara Department of Materials Science, Graduate School of Science and Engineering, Tokyo Institute of Technology, 2-12-1-H61 Oh-okayama, Meguro-ku, Tokyo 152-8551, Japan, e-mail: skoshi@cms.titech.ac.jp

Shigemi Mizukami WPI Advanced Institute for Materials Research, Tohoku University, 2-1-1 Katahira, Aoba-ku, Sendai 980-8577, Japan, e-mail: mizukami@wpi-aimr.tohoku.ac.jp

Koki Takanashi Magnetic Materials Laboratory, Institute for Materials Research, Tohoku University, 2-1-1 Katahira, Aoba-ku, Sendai 980-8577, Japan, e-mail: koki@imr.tohoku.ac.jp

Yoichi Takanishi Division of Physics and Astronomy, Graduate School of Science, Kyoto University, Kitashirakawa-oiwakecho, Sakyo, Kyoto 606-8502, Japan, e-mail: ytakanis@scphys.kyoto-u.ac.jp

Chapter 1
Quantum Structures of Advanced Materials

Yoshinobu Aoyagi

Abstract In this chapter the fundamental optical properties of low dimensional semiconductor materials like quantum well, wire and dot will be explained from analytical points of view in addition to real application of those low dimensional materials to devices. At the beginning the analytical method of hetero junction will be explained which is a base of quantum structures. After that the density of states of low dimensional material will be calculated for free electron in materials which is important to understand the optical properties of real devices. The optical properties of each low dimensional material will be explained and readers can understand what the specific optical nature of each low dimensional material is. Finally in this section the real application of low dimensional material to laser will be introduced and advantage of the low dimensionality will be explained. Since many references are cited, readers can understand deeply the parts which are interested in through the references.

1.1 Quantum Structures of Advanced Materials

1.1.1 Introduction

When a super-lattice was proposed by Prof. L. Ezaki for the first time, we learnt that artificial nano structure material like the super-lattice showed various interesting properties explained only by quantum mechanics. Some parts of properties were not obtained in natural world. We obtained wisdom not only that the band structure of the semiconductor and the metal was given in nature, but also that was able to control freely. Now the understanding of optical properties of the artificial new materials

---

Y. Aoyagi (✉)
Global Innovation Research Organization, Ritsumeikan University,
Noji-Higashi, Kusatsu-shi 1-1-1, Shiga, 525-8577 Japan
e-mail: aoyagi@fc.ritsumei.ac.jp

Y. Aoyagi and K. Kajikawa (eds.), *Optical Properties of Advanced Materials*,
Springer Series in Materials Science 168, DOI: 10.1007/978-3-642-33527-3_1,

like the super-lattice, the quantum wire, and the quantum dot is very important. This chapter treats the event related to optical properties of the artificial nano-structured materials.

1.1.2 Heterostructure

For easy understanding the hetero-structure of p-GaAs and n-Ge will be discussed. The band diagram of each material is shown in Fig. 1.1.

For the calculation going underneath, parameters shown in Table 1.1 are used. In this figure the Fermi level, and energy of valence band and conduction band for p-GaAs and n-Ge are shown E_{fp}, E_{fn}, E_{vp}, E_{vn}, E_{cp}, E_{cn}, respectively. When forming the hetero-structure the Fermi level of each material should be coincident. So, the band should bend and notch and/or jump of band structure at the interface should appear. First of all these values are calculated (Table 1.1).

$$E_{fp} = (\chi_{GaAs} + E_g(GaAs)) - \delta_{GaAs}) = 5.72\,eV \tag{1.1}$$

$$E_{fn} = \chi_{Ge} + \delta_{Ge} = 4.23\,eV \tag{1.2}$$

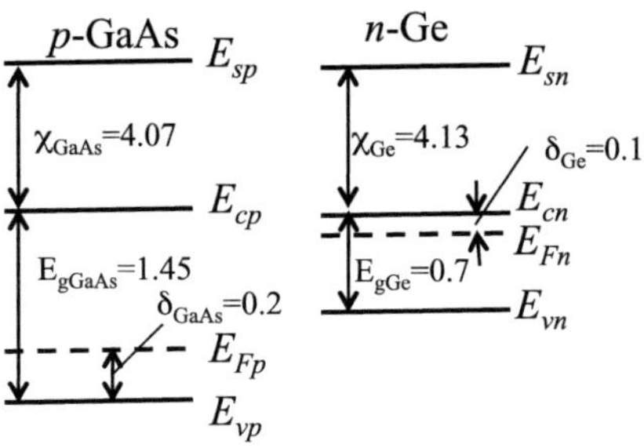

Fig. 1.1 Band structure of p-GaAs and n-Ge before formation of the hetero structure

Table 1.1 Fundamental parameters for the calculation of band diagram of p-GaAs/n-Ge hetero structure

	p-GaAs	n-Ge
Band gap (E_g)	1.45 eV	0.7 eV
Electron affinity χ	4.07 eV	4.13 eV
Donor density N_d		$1 \times 10^{16}\,cm^{-3}$
Acceptor density N_a	$3 \times 10^{16}\,cm^{-3}$	
$E_f - E_v = \delta_{GaAs}$	0.2 eV	
$E_f - E_v = \delta_{Ge}$		$0.1\,eV$
Relative dielectric constant ϵ	11.5	16

As mentioned above, since the Fermi level of each material after forming the hetero-structure should be coincident, the a voltage appears at both sides of hetero-structure due to the charge transfer and the vacuum level should be bended with a value of V_{Dn} and V_{Dp} due to this voltage. The sum of the band $V_{Dn}+V_{Dp}$ is calculated to be

$$E_{fp}-E_{fn} = (\chi_{Ge}+E_g(Ge)-\delta_{Ge})-(\chi_{GaAs}+\delta_{GaAs}) = V_{Dn}+V_{Dp} = 1.45\,\text{eV} \quad (1.3)$$

If we assume the charge appears only in the region of X_n and X_p near the interface for simple calculation,

$$X_n/X_p = N_A/N_D = 3 \quad (1.4)$$

from charge conservation low. From Poisson's equation

$$d^2V/dx^2 = -\rho/\epsilon \quad (1.5)$$

V_{Dn} and V_{Dp} can be easily calculated as follows;

$$V_{Dn} = N_D X_n^2/(2\epsilon_{Ge}) \quad (1.6)$$

$$V_{Dp} = N_A X_p^2/(2\epsilon_{GaAs}) \quad (1.7)$$

$$V_{Dn}/V_{Dp} = (N_{GaAs}/N_{Ge}) \cdot (\epsilon_{GaAs}/\epsilon_{Ge}) = 2.16 \quad (1.8)$$

So,

$$V_{Dn} = 2.16 V_{Dp} \quad (1.9)$$

In this case the band discontinuity is named to be the energy spike ΔE_v. The energy kink or the notch and the jump of band structure ΔE_c also appear. The each value is calculated to be

$$\delta E_c = \chi_{GaAs} - \chi_{Ge} = -0.06\,\text{eV} \quad (1.10)$$

$$\delta E_v = (E_{gGe} - E_{gGaAs}) - (\chi_{GaAs} - \chi_{Ge}) = -0.69\,\text{eV} \quad (1.11)$$

So,

$$\delta E_c + \Delta E_v = E_{gGe} - E_{gGaAs}. \quad (1.12)$$

According to these calculations the band diagram of this p-GaAs and n-Ge is shown in Fig. 1.2. The details will be understood by referring the article [1].

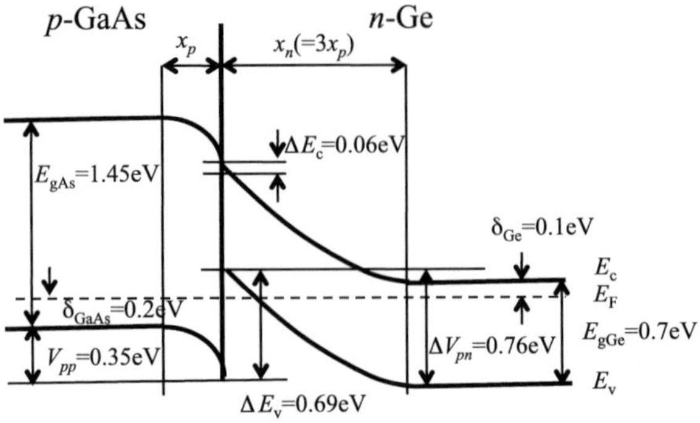

Fig. 1.2 Band structure of p-GaAs and n-Ge hetero-structure

1.2 Quantum Effects

1.2.1 Electron State in Semiconductor Quantum Well

We can make a hetero-structure of semiconductor by stacking different composite material of A and B like B/A/B, for example, AlGaAs/GaAs/AlGaAs as shown in Fig. 1.3. The band gap of material A and B is different. If the band gap of A is smalller than that of B we can construct quantum well structure shown in Fig. 1.3, and electron can be confined in the quantum well.

1.2.2 Energy State of Electron in Infinite Depth Quantum Well

Here, the rectangular well type potential problem shown in the semiconductor hetero-structure is taken up. First of all, let's evaluate the energy eigen-value of the well of infinite depth with an easy handling.

When the thickness of the well (the direction of x) is assumed to be L, and the starting point is taken at the center of the well, the potential structure is given as shown in Fig. 1.4. When width L is shorter than the de Broglie wavelength of the electron, the effect of quantum mechanics becomes strongly visible. Therefore, the well is especially called "Quantum well". Here, the center of the well was assumed to be a coordinate origin, and the energy of potential in the well was assumed to be 0. At this time, Schrödinger equation in the well is given by (1.13). As the boundary condition, because the electron cannot exist outside of the well, wave function given by Eq. 1.14 is used. Because the outside in this well is a barrier for the existence of electron, it is especially called barrier layer.

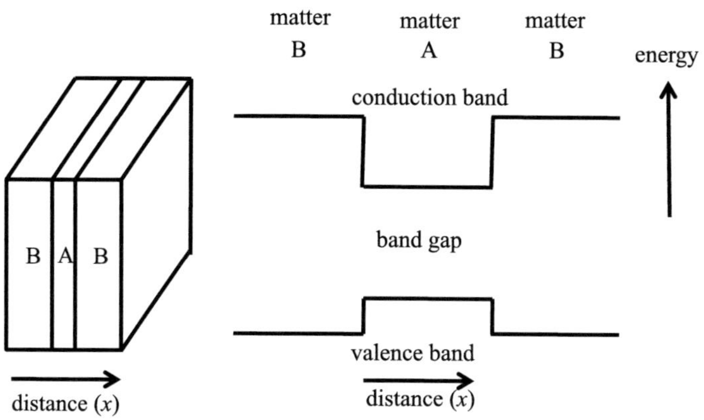

Fig. 1.3 Typical hetero-structure and the band structure

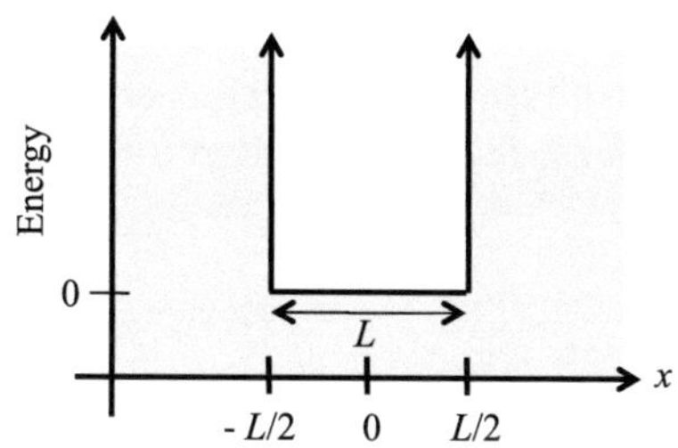

Fig. 1.4 Potential structure of quantum well with barrier of infinity of depth

$$-\frac{\hbar^2}{2m}\frac{d^2\psi(x)}{dx^2} = E\psi(x) \qquad (|x| < \frac{L}{2}) \tag{1.13}$$

$$\psi(x) = 0 \qquad (|x| \geq \frac{L}{2}) \tag{1.14}$$

Energy related team in Eq. 1.13 is expressed as

$$k = \sqrt{\frac{2mE}{\hbar^2}} \tag{1.15}$$

It is possible to show by putting it into Eq. 1.13 as follows.

$$\frac{d^2\psi(x)}{dx^2} = -k^2\psi(x) \tag{1.16}$$

It is understood that Schrödinger equation shown above has the shape of an easy second order differential equation. The solution of Eq. 1.16 is given as,

$$\psi(x) = C_1\cos(kx) \tag{1.17}$$

$$\psi(x) = C_2\sin(kx) \tag{1.18}$$

Here, C_1 and C_2 are the integration constants, respectively.

For satisfying a boundary condition of Eq. 1.17, an equation $\cos(kL/2) = 0$ should be satisfied. So as to fill this equation

$$k\frac{L}{2} = \frac{\pi}{2}(2q - 1) \qquad (q = 1, 2, 3, \ldots) \tag{1.19}$$

should be satisfied. For the boundary condition of the Eq. 1.18, it is necessary to fill the equation

$$k\frac{L}{2} = \frac{\pi}{2}2q \qquad (q = 1, 2, 3, \ldots) \tag{1.20}$$

when the expression 1.19 and the expression 1.20 are brought together, it becomes

$$k = \frac{\pi}{L}n \qquad (q = 1, 2, 3, \ldots) \tag{1.21}$$

From Eqs. 1.15 and 1.21, energy state E in the quantum well corresponding to each eigenstate n is obtained as follows;

$$E_n = \frac{\hbar^2}{2m}\left(\frac{\pi}{L}\right)^2 n^2 \tag{1.22}$$

When n is odd number, the wave function in the quantum well has even symmetry as shown in Eq. 1.17 and when n is even number, the wave function has odd symmetry shown in Eq. 1.18, that is,

$$\psi(x) = \sqrt{\frac{2}{L}}\cos\left(\frac{n\pi}{L}x\right) \qquad n = \text{odd number} \tag{1.23}$$

$$\psi(x) = \sqrt{\frac{2}{L}}\sin\left(\frac{n\pi}{L}x\right) \qquad n = \text{even number} \tag{1.24}$$

Integration constant C of the expression 1.23 and 1.24 is a normalized constant. The integrated value of the second power of the wave function in whole space must become one, because the total probability of finding electron in the whole space should be one.

1.2.3 Energy State in a Limited Depth Quantum Well

Next, let's take up the problem when the depth of a rectangular square well potential has a limited value of V. The thickness of the well is assumed to be L as shown in Fig. 1.5 and the starting point is taken at the center of the well. Electron energy E in the well is computed as follows. The effective mass of the electron in the barrier layer and the well layer is assumed to be m_B and m_w respectively now.

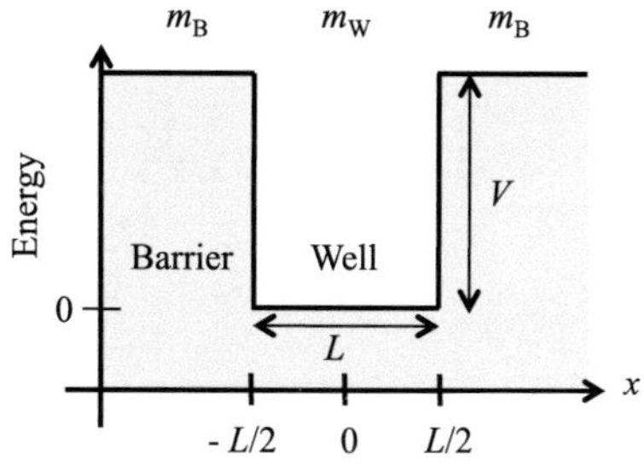

Fig. 1.5 Potential structure of quantum well

Because the potential energy is 0 in the well, the Schrödinger equation is similar as Eq. 1.13 and the solution in the well is also similar. That is, Schrödinger equation is given by

$$-\frac{\hbar^2}{2m_W}\frac{d^2\psi(x)}{dx^2} = E\psi(x) \qquad (|x| < \frac{L}{2}) \tag{1.25}$$

The solution is

$$\psi(x) = C_3 \cos(k_0 x) \tag{1.26}$$

$$\psi(x) = C_4 \sin(k_0 x) \tag{1.27}$$

Here,

$$k_0 = \sqrt{\frac{2m_W E}{\hbar^2}} \tag{1.28}$$

Here, C_3, and C_4 are the integration constants, respectively. It becomes from the shape of the solution that in case of quantum number is odd, the wave function is shown by a symmetric expression shown in Eq. 1.17 and the case of even, the function is asymmetric expression 1.18.

On the other hand, outside the well ($|x| \geq L/2$), the Schrödinger equation is given by

$$-\frac{\hbar^2}{2m_W}\frac{d^2\psi(x)}{dx^2} + V\psi(x) = E\psi(x) \tag{1.29}$$

In consideration of $E < V$,

$$k = \sqrt{\frac{2m_B(V-E)}{\hbar^2}} \tag{1.30}$$

The differential equation can be rewritten as

$$\frac{d^2\psi(x)}{d^2} = k^2\psi(x) \tag{1.31}$$

The solution becomes

$$\psi(x) = C_5 \exp(-kx) \qquad (x \geq \frac{L}{2}) \tag{1.32}$$

$$\psi(x) = C_6 \exp(kx) \qquad (x \leq -\frac{L}{2}) \tag{1.33}$$

Here, the wave function $\psi(x)$ should be zero at $x \to \pm\infty$ from the physics implication. So, Eqs. 1.32, 1.33 are obtained by excluding the solution which diverse at the infinite value of x. Here, C_5 and C_6 are the integration constants, respectively. Depending on the symmetry of the solution, the integration constant has a relation of $C_5 = C_6$ (symmetric case) and $C_5 = -C_6$ (asymmetric case).

From the continuous condition of existence probability and probability density flow of electron at any place, wave function $\psi(x)$ and $(1/m)(d\psi(x)/dx)$ should be continuous in any place. By putting the wave function inside the well and outside the well to be $\Psi_{\mathrm{w}}(x)$ and $\Psi_{\mathrm{B}}(x)$

$$(1/m_{\mathrm{w}})(d\Psi_{\mathrm{w}}/dx)(-L/2) = (1/m_{\mathrm{B}})(d\Psi_{\mathrm{B}}/dx)(-L/2) \tag{1.34}$$

$$(1/m_{\mathrm{w}})(d\Psi_{\mathrm{w}}/dx)(L/2) = (1/m_{\mathrm{B}})(d\Psi_{\mathrm{B}}/dx)(L/2) \tag{1.35}$$

By applying continuous condition at $x = L/2$ for symmetric wave function 1.26 in the well (same in case of $-L/2$) we obtain following equation.

$$C_3 \cos\left(\frac{kL}{2}\right) = C_5 \exp\left(-\frac{k_0 L}{2}\right) \tag{1.36}$$

$$-C_3\left(\frac{k}{m_{\mathrm{w}}}\right) \sin\left(\frac{kL}{2}\right) = -C_5\left(\frac{k_0}{m_{\mathrm{B}}}\right) \exp\left(-\frac{k_0 L}{2}\right) \tag{1.37}$$

By dividing Eq. 1.36 by Eq. 1.37, the integration constant C_3 and C_5 can be deleted, and the energy eigenvalue is obtained.

$$\left(\frac{k}{m_{\mathrm{w}}}\right) \tan\left(\frac{k_0 L}{2}\right) = \frac{k}{m_{\mathrm{B}}} \tag{1.38}$$

Because k_0 and k are functions of energy E, from the eigenvalue equation

$$\tan\left(\frac{k_0 L}{2}\right) = \frac{k}{k_0}\frac{m_{\mathrm{w}}}{m_{\mathrm{B}}} \tag{1.39}$$

the eigenvalue E can be numerically calculated. Similarly, for the asymmetric wave 1.27 the following eigen equation is obtained by applying the continuous condition at $x = L/2$.

$$\cot\left(\frac{kL}{2}\right) = -\frac{k_0}{k}\frac{m_{\mathrm{w}}}{m_{\mathrm{B}}} \tag{1.40}$$

In other word, to determine intrinsic energy E_n of the quantum well of limited depth corresponding to eigenstate n, will only have to solve eigen equation 1.39 to the one of the odd number and to solve 1.40 for the one of the even number. In this case, wave function ψ_n to the following eigenvalue should satisfy normalization condition of each n of the wave function. That is,

$$\int_{-\infty}^{\infty} \psi_n^* \psi_n dx = \int_{-\infty}^{-L/2} \psi_{-,n}^* \psi_{-,n} dx + \int_{-L/2}^{L/2} \psi_{0,n}^* \psi_{0,n} dx + \int_{L/2}^{\infty} \psi_{+,n}^* \psi_{+,n} dx = 1 \tag{1.41}$$

By using this condition when n is an odd number

$$\psi_{0,n}(x) = 2A_{0,n} \cos(k_{0,n}x) \qquad (-L/2 < x \leq L/2) \tag{1.42}$$

$$\psi_{+,n}(x) = 2A_{0,n} \cos(k_{0,n}L/2) \exp^{-k_n x} \qquad (x \geq L/2) \tag{1.43}$$

$$\psi_{-,n}(x) = 2A_{0,n} \cos(k_{0,n}L/2) \exp^{k_n x} \qquad (x \leq -L/2) \tag{1.44}$$

where

$$A_{0,n} = \left(\frac{e^{-k_n L/2}}{k_n} \cos^2(k_{0,n}L/2) + \frac{1}{2k_{0,n}} \sin(2k_{0,n}L/2) + L/2\right)^{-\frac{1}{2}} \qquad (n = 1, 3, 5, \ldots) \tag{1.45}$$

$\psi_{-,n}$ and $\psi_{+,n}$ show the following peculiar wave function toward negative and positive direction, respectively from a center of the well. In case of even number n, solve the wave function by each of the reader.

When V is infinite potential

$$2A_{0,n} = \sqrt{2/L} \tag{1.46}$$

And $\psi_{-,n} = \psi_{+,n} = 0$ at the wall. Moreover,

$$\psi_{0,n} = \sqrt{\frac{2}{L}} \cos\left(\frac{n\pi}{L}x\right)_{n=1,3,5\cdots} \qquad n = 1, 3, 5, \ldots \tag{1.47}$$

$$\psi_{0,n} = i\sqrt{\frac{2}{L}} \sin\left(\frac{n\pi}{L}x\right)_{n=2,4,6\cdots} \qquad n = 2, 4, 6, \ldots \tag{1.48}$$

Equations 1.47 and 1.48 agree with the solution solved by infinite potential.

The technique to find a solution of eigen energy is commonly done using graphical analysis, which is adopted by many textbook. However, a strict analysis should be done using computer analysis. I will introduce new computer analysis method here according to a method proposed by Prof. Misaichi Takeuchi, Ritsumeikan University.

1.2.4 Method and Idea of the Calculation

The energy of an eigenvalue of ground state ($n = 1$) of the quantum well with the width L and the infinite depth is defined to be E_1. (This value has been given in Eq. 1.22 already with $n = 1$). Thereafter, energy of eigenvalue E for the well with a limited depth can show with the value of $E_1\left(=\frac{\hbar^2}{2m_w}\left(\frac{\pi}{L}\right)^2\right)$ as a unit and a. So,

$$E = E_1 \times a^2 = \frac{\hbar^2}{2m_w}\left(\frac{\pi}{L}\right)^2 \times a^2 \tag{1.49}$$

Solving the eigenvalue problem for well with finite depth is equivalent to solving the equation of a. In case of infinite well case the value of a takes integer $(1, 2, 3, 4, \ldots)$ but in the case of finite value of well depth a is not an integer value.

By the way, what is numerical value a? What is the physics implication of a? From the expression 1.28, k_0 can be rewritten as

$$k_0 = \frac{\pi_0}{L}a \tag{1.50}$$

Wave function 1.26 is also rewritten as

$$\psi(x) \propto \cos\left(\frac{\pi}{L}ax\right) \tag{1.51}$$

If a is one, it is the case of the infinite depth quantum well. In case of the finite depth quantum well, the condition of a becomes $0 < a < 1$. This becomes clear if you think about the shape of the wave function.

As shown in Fig. 1.6, for the infinite depth quantum well it is necessary to become $\psi(x) = 0$ at $x = \pm L/2$. However, the wave function extends into the barrier as an evanescent wave for the limited depth quantum well. It becomes the one showing the expression 1.32 and 1.33 as expressed in a term of e^{-kx}, e^{kx}. To keep continuity of wave function between cos part in the well and an evanescent part in the barrier,

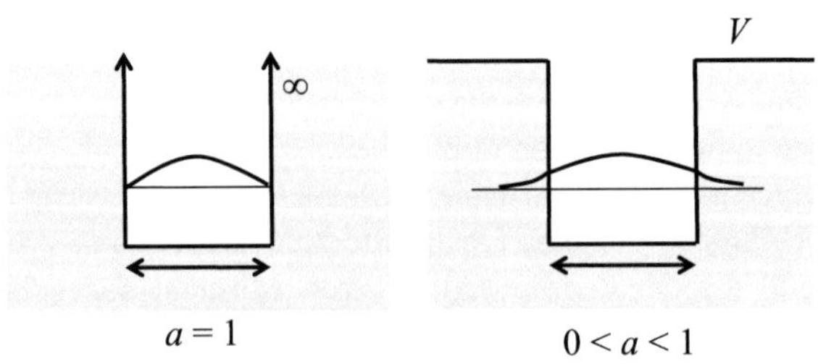

Fig. 1.6 Physics implication of a is shown. The *left* the case of the infinite potential of the well. The *right* the case of the limited potential well. The position of the center of well is assigned to be 0 and the position of the well is assigned to be $x = \pm L/2$

$0 < a < 1$ condition is required. The value of a corresponds to a degree of the penetration of wave function into the barrier.

As a result, as for eigenvalue Eq. 1.39, $1 < a < 0$ is a condition of obtaining the lowest energy eigenvalue of the quantum well of limited depth well. The number of the quanta in this case is still $n = 1$. In this case the Eq. 1.39 is rewritten as

$$\tan\left(\frac{k_0 L}{2}\right) = \frac{k}{k_0} \cdot \frac{m_W}{m_B} = \sqrt{\frac{V-E}{E}} \cdot \sqrt{\frac{m_W}{m_B}} \tag{1.52}$$

V can be expressed using a unit of E_1 as follows;

$$V = E_1 \times V' \tag{1.53}$$

So, eigenvalue equation is rewritten using a as follows

$$\tan\left(\frac{\pi}{2}a\right) = \sqrt{\frac{V'-a^2}{a^2}} \cdot \sqrt{\frac{m_W}{m_B}} \tag{1.54}$$

That is,

$$\tan\left(\frac{\pi}{2}a\right) - \sqrt{\frac{V'-a^2}{a^2}} \cdot \sqrt{\frac{m_W}{m_B}} = 0 \tag{1.55}$$

The solution of this equation can be found using simple computer calculation under the condition of $0 < a < 1$. An actual energy eigenvalue is obtained from the expression 1.49 using an obtained value of a. In case of $n = 3$, the solution can be obtained under the condition of $2 < a < 3$. Equation 1.55 is the equation for $n =$ odd number.

When n is an even number, it is noted that the eigen equation becomes an expression 1.40 and is rewritten as;

$$\cot\left(\frac{\pi}{2}a\right) = -\sqrt{\frac{V'-a^2}{a^2}} \cdot \sqrt{\frac{m_W}{m_B}} \tag{1.56}$$

That is,

$$\cot\left(\frac{\pi}{2}a\right) + \sqrt{\frac{V'-a^2}{a^2}} \cdot \sqrt{\frac{m_W}{m_B}} = 0 \tag{1.57}$$

By solving this equation, the eigenvalue for even number of n can be obtained by using computer analysis just same as to the case of odd number of n. In case of $n = 2$, the condition to solve the above equation is $1 < a < 2$, and for $n = 4$, $3 < a < 4$. From this calculation it will be clear that eigen state at least more than 1 exist in any quantum well with symmetric potential well shown in following Fig. 1.7.

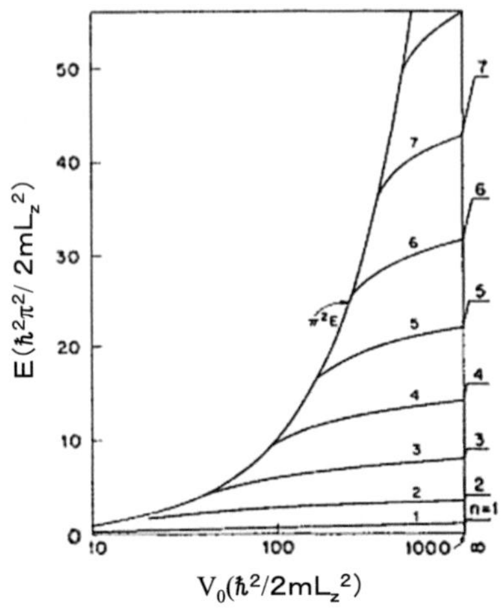

Fig. 1.7 Calculated normalized energy levels of an electron in a symmetrical rectangular potential well of depth V [2]

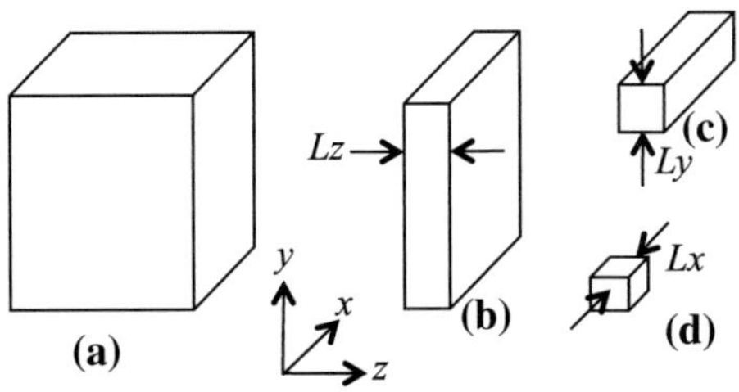

Fig. 1.8 Schematic structures of various quantum structures. **a** Balk, **b** quantum well, **c** quantum wire, **d** quantum dot

The reader is recommended to calculate the eigen energy using this method by a computer. Figure 1.7 shows the calculated normalized eigen states as a function of potential barrier height for each n. As shown in this figure, at least more than one eigen state exists in any kind of potential well.

1.2.5 Density of States

The density of state is a number of electrons that can be occupied in energy dE of E and $E + dE$ for the electron with energy E. Let's calculate the case of a free electron as a simplified example. How is the density of state for three, two, one and zero dimensional material shown in Fig. 1.8 calculated?

Three dimension system (bulk) k_x, k_y, k_z, and m^* are assumed to be momentum of electron for x, y and z direction and the effective mass, respectively. The energy of a free electron in three dimensions is given by

$$E = \frac{\hbar^2}{2m^*}(k_x^2 + k_y^2 + k_z^2) \tag{1.58}$$

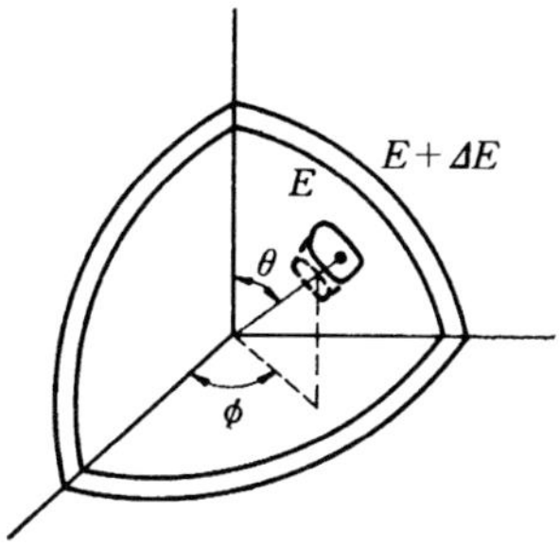

Fig. 1.9 Polar coordinates display of a basic element in k space

In one dimensional case a unit size of k space in which one electron occupy is $2\pi/L$ for the material with a size of L. So, for three dimensional case the size is given by $(2\pi/L)^3$.

If we consider the spin state in an electron, two electrons come to enter in this size. When the polar coordinate display shown in Fig. 1.9 is taken, the size of microelement dV_k given by $dk_x \times dk_y \times dk_z$ can be expressed using polar coordinates of θ and ϕ. In the polar coordinates the microelement dV_k is given by

$$dV_k = k^2 \sin\theta d\theta d\phi dk \tag{1.59}$$

The number of states of electron with an energy between E and $E + dE$ is given by integrating dV_k around all of the k space for electron with the energy between E and $E + dE$, that is, integrating the value for θ and ϕ, with a consideration of spin state of 2 for each unit and dividing the unit size of k space for one electron as follows;

$$\begin{aligned}\rho(E)dE &= \left(\frac{2\pi}{L}\right)^{-3} \int 2dV_k = \left(\frac{2\pi}{L}\right)^{-3} \int 2k^2 \sin\theta d\theta\, d\phi dk \\ &= \left(\frac{2\pi}{L}\right)^{-3} \int 2\frac{k^2 \sin\theta d\theta d\phi}{dE/dK} dE\end{aligned} \tag{1.60}$$

$k^2 \sin\theta d\theta d\phi$ generally becomes an area of elemental energy unit on an energy constant surface in the Brillouin zone, although details are not described here. Moreover, the above expression is universal for any energy surface like not only a hyperbolic surface of a free electron but also for a complex energy curved surface. The density of state can obtain by calculating dE/dK of the denominator with a complex energy curved surface in general. For example, when the energy surface is an oval, following equation for $\mathrm{d}E/dk$ is given.

$$\frac{dE}{dk} = \left(\left(\frac{\partial E}{\partial k_x}\right)^2 + \left(\frac{\partial E}{\partial k_y}\right)^2 + \left(\frac{\partial E}{\partial k_z}\right)^2\right)^{-\frac{1}{2}} \tag{1.61}$$

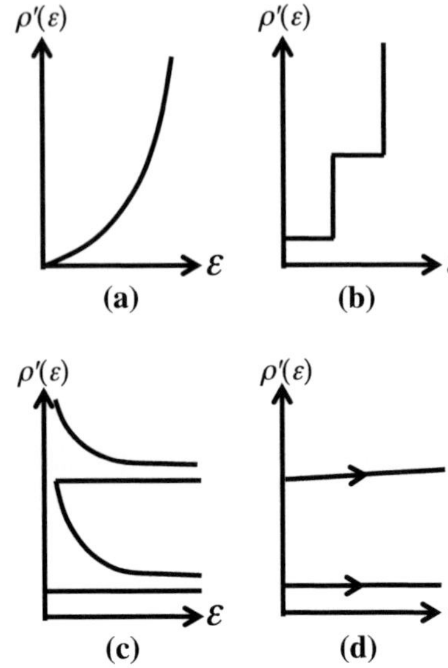

Fig. 1.10 Density of state in quantized structure in various dimensions. **a** Bulk, **b** quantum well, **c** quantum wire, and **d** quantum dot

Here, the simplest case of free electron, which has a parabolic energy surface, is taken into account. Moreover, let's use a handy method for a free electron to determine the density of state, though the way is not general.

In this case, total number of the states with the momentum from 0 to k takes even the spin into consideration is given by

$$\frac{2}{\left(\frac{2\pi}{L}\right)^3} = \int_0^k 4\pi k^2 dk = \frac{V}{3\pi^2}\left(\frac{2m^* E}{\hbar^2}\right)^{\frac{3}{2}} \tag{1.62}$$

by taking the unit volume $V = L^3$.

Because the density of state is the number of state occupies between E and $E+dE$, the value can be obtained by differentiating the above expression

$$\rho(E)dE = \frac{1}{3\pi^3}\left(\frac{2m^*}{\hbar^2}\right)^{\frac{3}{2}} \times \frac{3}{2} \times \sqrt{E}dE = \frac{1}{2\pi^3}\left(\frac{2m^*}{\hbar^2}\right)^{\frac{3}{2}} \times \sqrt{E}dE \tag{1.63}$$

If this is written in figure, it shows in Fig. 1.10a. Remember the calculation of the density of state is essential for evaluating the electron number in conduction band in semiconductor. In that case the equation of the density of state is same as the equation written above.

The density of state for two dimensions (quantum well) In this case the energy of electron is given by

$$E_n = \frac{\hbar^2}{2m^*}(k_x^2 + k_y^2) + E_{n_z}, \tag{1.64}$$

where

$$E_{n_z} = \frac{\hbar^2}{2m^*}\left(\frac{\pi n_z}{L_z}\right)^2 \tag{1.65}$$

So, total density of state in this case is given by,

$$\frac{2}{\left(\frac{2\pi}{L}\right)^2}\int_0^k 2\pi k dk = \frac{L^2}{2\pi}k^2 = \frac{L^2}{2\pi}(E_n - E_{n_z})\frac{2m^2}{\hbar^2} \tag{1.66}$$

Then,

$$\rho(E - E_n)dE = \frac{Lm^*}{\pi\hbar^2}dE \tag{1.67}$$

By considering all energy states existing in the well, the total density of state in quantum well is given by

$$\rho(E - E_n)dE = \sum_{n_z}\rho(E - E_{n_z})dE \tag{1.68}$$

So, the density of state is given by

$$\rho(E - E_n) = \sum_{n_z}\frac{Lm^*}{\pi\hbar^2}\Theta(E - E_{n_z}) \tag{1.69}$$

where Θ is a step function. This result is shown in Fig. 1.10b.

The density of state for one dimensions (quantum wire) The energy of electron in one dimensional case is given by

$$E = \frac{\hbar^2}{2m^*}k_x^2 + E_{n_y} + E_{n_z} \tag{1.70}$$

where

$$E_{n_y} = \frac{\hbar^2}{2m^*}\left(\frac{\pi n_y}{L_y}\right)^2 \tag{1.71}$$

$$E_{n_z} = \frac{\hbar^2}{2m^*}\left(\frac{\pi n_z}{L_z}\right)^2 \tag{1.72}$$

In this case the total number of the states of k space is given by

$$\frac{2}{\frac{2\pi}{L}}\int_{-k}^{k} dk = \frac{2L}{\pi}k = \frac{2L}{\pi}\left(\frac{2m^*}{\hbar^2}\right)^{\frac{1}{2}}\sqrt{E - E_{n_y} - E_{n_z}} \tag{1.73}$$

So, the density of the state in this case is given by

$$\rho(E)dE = \left(\frac{2m^*}{\hbar^2}\right)^{\frac{1}{2}} \frac{2L}{\pi} \frac{1}{2} \frac{1}{\sqrt{E - E_{n_y} - E_{n_z}}} = \frac{\sqrt{2m^*}L}{\pi\hbar} \frac{1}{\sqrt{E - E_{n_y} - E_{n_z}}} \tag{1.74}$$

By considering all possible states in the quantum wire which is quantized is given by

$$\rho(E)dE = \sum \rho(E - E_{n_y} - E_{n_z})dE \tag{1.75}$$

So, the density of states is given by,

$$\rho(E) = \frac{\sqrt{2m^*}L}{\pi\hbar} \sum_{n_y,n_z} \frac{1}{\sqrt{E - E_{n_y} - E_{n_z}}} \tag{1.76}$$

This case is shown in Fig. 1.10c.

Zero dimensional case In this case the energy of electron is given by

$$E = E_{n_x} + E_{n_y} + E_{n_z} \tag{1.77}$$

And the density of states in this case is given by

$$\rho(E) = 2 \sum_{n_x,n_y,n_z} \delta(E - E_{n_x} - E_{n_y} - E_{n_z}). \tag{1.78}$$

If this is written in the figure, it becomes Fig. 1.10d. It is assumed that the effective mass with the barrier and the quantum well, wire, and dot are same in this figure.

1.3 Optical Properties in Semiconductor Quantum Structures

1.3.1 Optical Properties of Quantum Well

When n_e and n_h are the eigenvalue of the quantum well for electron and hole, respectively, the energy of eigen states of conduction band and valence band E_e and E_h, respectively are given by

$$E_\mathrm{c} = \frac{\pi\hbar^2}{2m_\mathrm{e}^*}(k_x^2 + k_y^2) + E_\mathrm{e}(n_\mathrm{e}) + E_\mathrm{g} \tag{1.79}$$

$$E_\mathrm{h} = \frac{\pi\hbar^2}{2m_\mathrm{h}^*}(k_x^2 + k_y^2) + E_\mathrm{h}(n_\mathrm{h}) \tag{1.80}$$

The vertical direction z is assumed to be a transverse direction of the quantum well plane of x and y direction, and when an envelope wave function of each electron and hole and the Bloch function are given by $\psi(z)$ and u, respectively, the wave function in z direction is given by

$$\phi_{\mathrm{ne}}(r) = \Psi_{\mathrm{ne}}(z)\exp^{i(k_x x + k_y y)} u_{\mathrm{c}} \tag{1.81}$$

$$\phi_{\mathrm{nh}}(r) = \Psi_{\mathrm{nh}}(z)\exp^{i(k_x x + k_y y)} u_{\mathrm{v}} \tag{1.82}$$

Where, u_{c} and u_{v} is a Bloch function of electron in conduction band and a Bloch function of hole in valence band. The transition matrix element for dipole moment is given by

$$< \phi_{\mathrm{ne}}|p|\phi_{\mathrm{nh}} >=< \psi_{\mathrm{ne}}(z)|\psi_{\mathrm{nh}}(z) >< u_{\mathrm{c}}|p|u_{\mathrm{v}} > \tag{1.83}$$

Because the transition of the top of valence band and the bottom of conduction band is permitted, $< u_{\mathrm{c}}|p|u_{\mathrm{v}} >$ is not zero. So,

$$< \psi_{\mathrm{ne}}(z)|\psi_{\mathrm{nh}}(z) > \neq 0 \tag{1.84}$$

Under the condition of that the optical transition between valence and conduction band really exists, the Eq. 1.83 is not zero. Therefore, the selection rule is given by the symmetry of the envelope wave function of eigen state n. So, for giving the condition of Eq. 1.84 the symmetry of valence band and conduction band should be same. So, the selection rule is given by

$$n_{\mathrm{ne}} = n_{\mathrm{nh}}(= n) \ \text{ or } \ \Delta n = n_{\mathrm{ne}} - n_{\mathrm{nh}} = 0 \tag{1.85}$$

Therefore, the transition energy is given by

$$\Delta E(n, k_x, k_y) = E_{\mathrm{g}} + E_{\mathrm{e}}(n) + E_{\mathrm{h}}(n) + \frac{\hbar^2}{2}\left(\frac{1}{m_{\mathrm{e}}^*} + \frac{1}{m_{\mathrm{h}}^*}\right)(k_x^2 + k_y^2) \tag{1.86}$$

Schematic diagram of the transition of the typical quantum well is shown in Fig. 1.11 in which heavy hole (hh) band and light hole (lh) band is included which comes from splitting of degeneracy of valence band by electron-orbit interaction. The observed actual spectrum, and the transition are shown in Fig. 1.12. The transition energy in this case is given by

$$\Delta E(n) = E_{\mathrm{g}} + E_{\mathrm{e}}(n) + E_{\mathrm{h}}(n) \tag{1.87}$$

In Fig. 1.13, the emission wavelength from AlGaAs/GaAs super lattice is shown as a function of well width. The solid line is calculated line according to the eigen states calculation. The experimental results are in good agreement with the result of the calculation. Moreover, the width of the line of the emission spectrum is narrow compared with the case of the bulk.

The density of state in the quantum well is stair like but that of balk is square root function. So, the width of the half-power of the emission spectrum of the quantum well becomes 0.7 times compared to that of the balk.

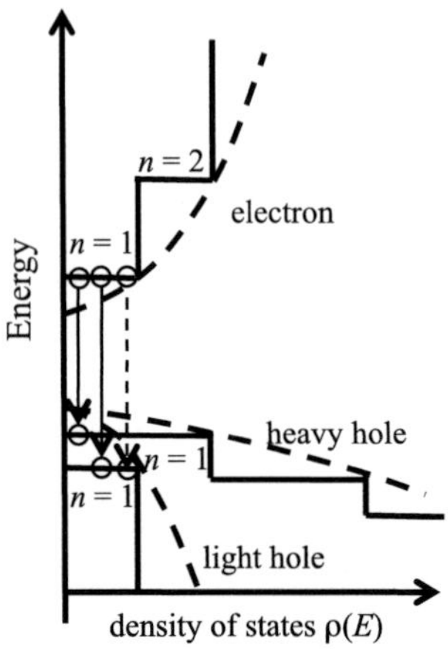

Fig. 1.11 Schematic energy diagram of quantum well for light hole and heavy hole with possible electrical transitions between conduction band and valence band

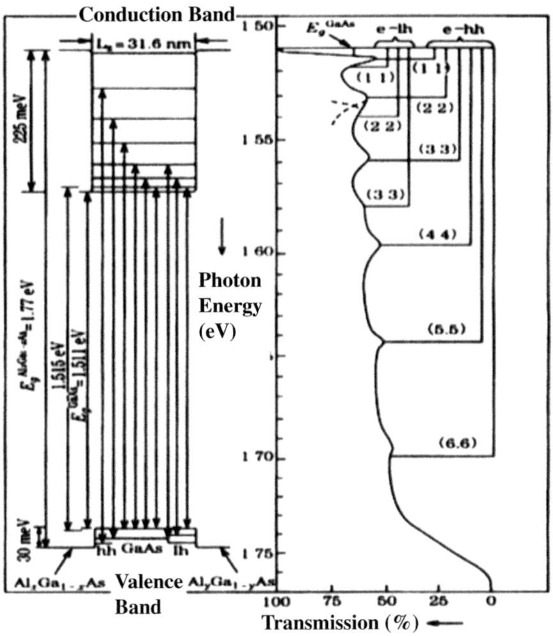

Fig. 1.12 Possible transition between quantum states and related absorption processes for GaAs quantum well with light hole and heavy hole [3]

1.3.2 Effect of Exciton in Semiconductor Quantum Well

The electron and hole forms a combined state called exciton at low temperature in balk. The binding energy at that time is given by

$$E_n = \frac{R_y^*}{n^2} \qquad (n = 1, 2, 3, \ldots) \tag{1.88}$$

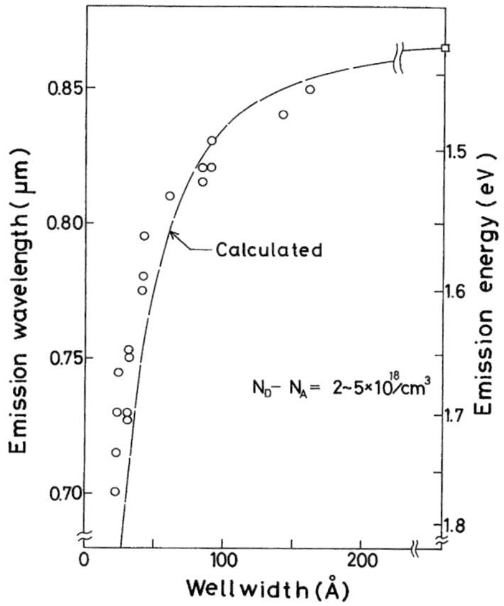

Fig. 1.13 Luminescence wavelength of quantum well versus quantum well layer thickness with experimental results and theoretical calculated curve. It is understood to show an extremely good agreement in theory and experimental results [4]

where, R_y^* is called exciton Rydberg constant and is given by

$$R_y^* = \frac{\mu e^4}{2\hbar^2 \epsilon^2} \tag{1.89}$$

ϵ is dielectric constant and μ is reduced effective mass of electron and hole

$$\mu^{-1} = m_{\mathrm{e}}^{-1} + m_{\mathrm{h}}^{-1} \tag{1.90}$$

At $n = 1$, $E_1 = R^*$, and this value is binding energy of the exciton. Then, how does the binding energy of the exciton become in the quantum well? [5]. The wave function of the exciton in complete two-dimensional system is given by solving a Schlödinger equation in which the coulomb interaction of electron and hole is considered. The Schlödinger equation is given by

$$-\frac{\hbar^2}{2\mu_\perp}\left(\frac{\partial^2}{\partial x^2} + \frac{\partial^2}{\partial y^2}\right)\Phi - \frac{e^2}{\epsilon\sqrt{x^2+y^2}}\Phi = E\Phi \tag{1.91}$$

When we use polar coordinate display, above equation is given by

$$\Phi(x, y) = \frac{1}{\sqrt{2\pi}} R(r) e^{im\phi} \tag{1.92}$$

$$\left[-\frac{\hbar^2}{2\mu}\left\{\frac{1}{r}\frac{d}{dr}\left(r\frac{d}{dr}\right) - \frac{m^2}{r^2}\right\} - \frac{e^2}{\epsilon r}\right]R(r) = ER(r), \tag{1.93}$$

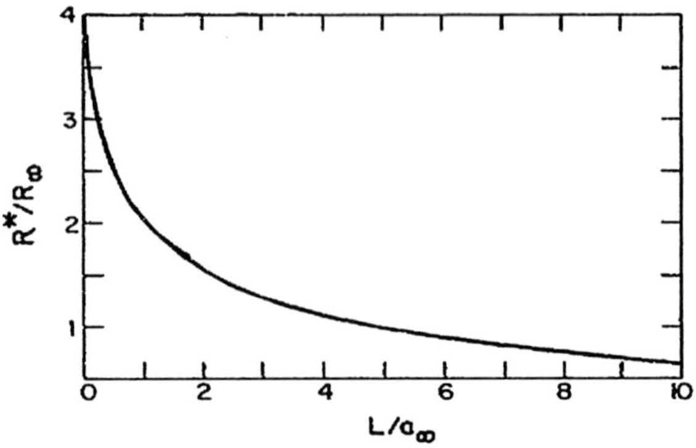

Fig. 1.14 Well thickness dependence of normalized binding energy of exciton of AlAs-GaAs-GaAlAs quantum well [5]

Fig. 1.15 Schematic diagram of density of states for balk and quantum well materials and the absorption spectrum for each bands

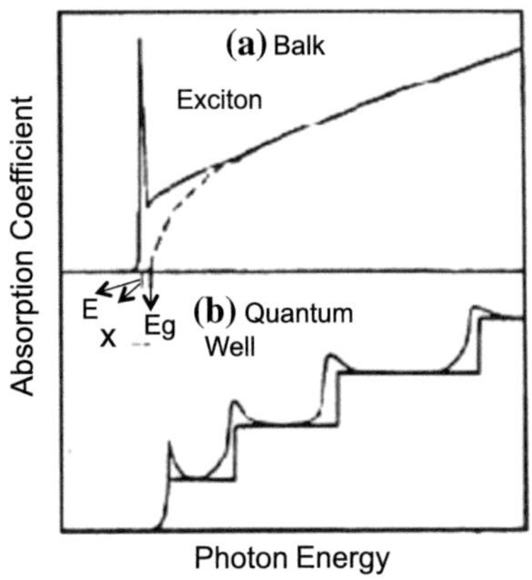

where m is an integer here ($m = 1, 2, 3, \ldots$). The solution of this equation, when the thickness of quantum well is extremely small, is given by

$$\phi_{nm}(r) = \left[\frac{(n-|m|)!}{\pi a^2 (n+\frac{1}{2})^3\{(n+|m|)!\}^3}\right] \times e^{-(\frac{1}{2})\rho}\rho^{|m|}L_{n+|m|}^{2|m|}(\rho)e^{im\varphi} \qquad (1.94)$$

Here, $L(\rho) = L_{n+|m|}^{2|m|}$, $|m| \leq n$, $\lambda^{-2} = -4w$, $w = E/R_y^*$, $a = \frac{h^2}{\mu e^2}$. $L(\rho)$ is polynomial of Laguerre. It can be understood that binding energy E_1 of exciton in a thin quantum well becomes four times larger than that of the bulk by solving 1.93 as an infinitely thin limiting case. In finite thickness case of quantum well, it is understood that the thicker it becomes as shown in Fig. 1.14 the smaller the binding energy and the energy approaches to the value of the bulk.

Figure 1.15 shows a density of state (dotted line) and the exciton absorption peak of the balk, which is shifted to lower energy side with an amount of binding energy

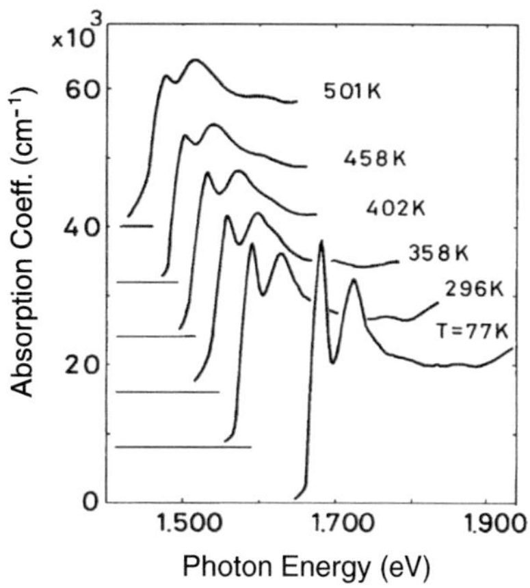

Fig. 1.16 Temperature dependence of the absorption spectrum observed in GaAs/AlGaAs quantum well. At low temperature, splitting of absorption from light hole and heavy hole to conduction band is observed [6]

of exciton. The lower figure shows a schematic diagram of density of state in the quantum well, and the absorption peak α in an absorption spectrum with exciton peaks.

In case of absorption of exciton it actually appears as two peaks because the degeneracy of a heavy hole and a light hole is resolved for the quantum well due to electron and orbit interaction, though here, the peak shows [6–8] as one peak for simplicity.

Figure 1.16 shows temperature dependence [6] of the photo-absorption spectrum of the exciton from GaAs/AlAs quantum well. Two peaks observed are originated from heavy hole exciton(hh) for $n = 1$, of which peak is lower energy side, and light hole exciton(lh) for $n = 1$, of which peak is higher energy side. The separation becomes unclear if the temperature rises and the wavelength shifts to the long wavelength, too. The person who wants to learn in details must refer to document [9–13].

1.3.3 Optical Properties of Quantum Wire

Research and application of optical properties of quantum wire have recently become active because high efficient light emission can be expected due to the high density of state and small number of defects in the crystal, especially in some kinds of quantum wire. Let's consider carbon nanotubes (hereafter CNT) as a typical example of the quantum wire here. CNT is a typical quantum wire and becomes semiconductor or metal depending on the chirality (how to roll the spiral). In Fig. 1.17 the typical band structure of metallic and semiconductor CNT, and the selection rule of optical transition between each band of each CNT are shown, respectively.

Since CNT is a typical one dimensional material, a strong selection rule between each band is defined. In case of semiconductor CNT the selection rule of the transition

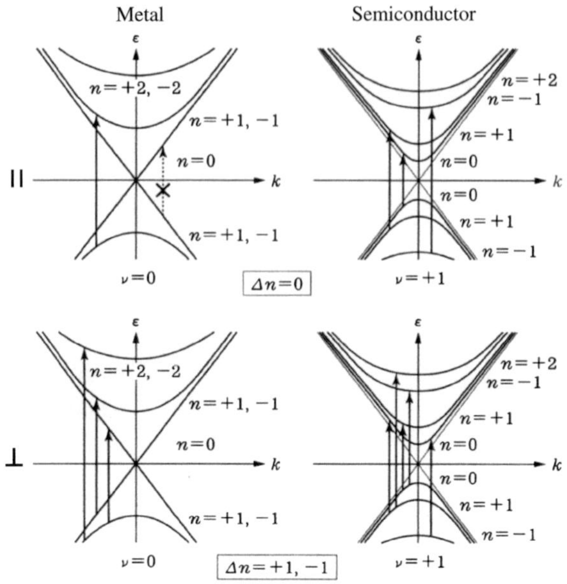

Fig. 1.17 Band structure and optical transition of metallic and semiconductive CNT. Figure in *left hand side* is a band structure of metallic CNT and the *right hand side* is that of semiconductive CNT. *Upper* figures show optical transition for the light with polarization parallel to the CNT and the *lower* is for the transverse polarization light [13]

is given by $\Delta n = 0$, in case of the direction of electric field is parallel to the direction of CNT. In case of the electric field is transverse to the direction of CNT, the selection rule becomes $\Delta n = \pm 1$. In case of metallic CNT, the band transition for $n = 0$ becomes forbidden exceptionally.

In Fig. 1.18 the density of state (DOS) [14] for (a) semiconductor CNT and (b) metallic CNT are shown. The density of state of graphene is written at the same time as dotted line for comparison. The spike of the density of state according to one dimensional structure of CNT appears in Fig. 1.18. The density of state in the band gap changes depending on chirality of CNT as shown in Fig. 1.18, and moreover, in case of semiconductor CNT no density of state in the gap is observed as shown in (a) but it exists for metallic CNT (b).

Luminescence from single CNT has come to be observed recently [15–21]. Figure 1.19 is SEM image of CNT that directly grown from the catalyst pillar without touching to the substrate to measure the optical property of single CNT. At beginning of the research on the luminescence from CNT, the semiconductor CNT had been expected to emit light. However, the luminescence from a single semiconductor CNT put on any substrate is not able to observe for a long time though luminescence from grass-like CNT was observed. Recently, it is understood that the strong luminescence is observed from CNT arranged so as not to come in contact with the substrate as shown in Fig. 1.19. The precise cause of the characteristic mentioned above has not

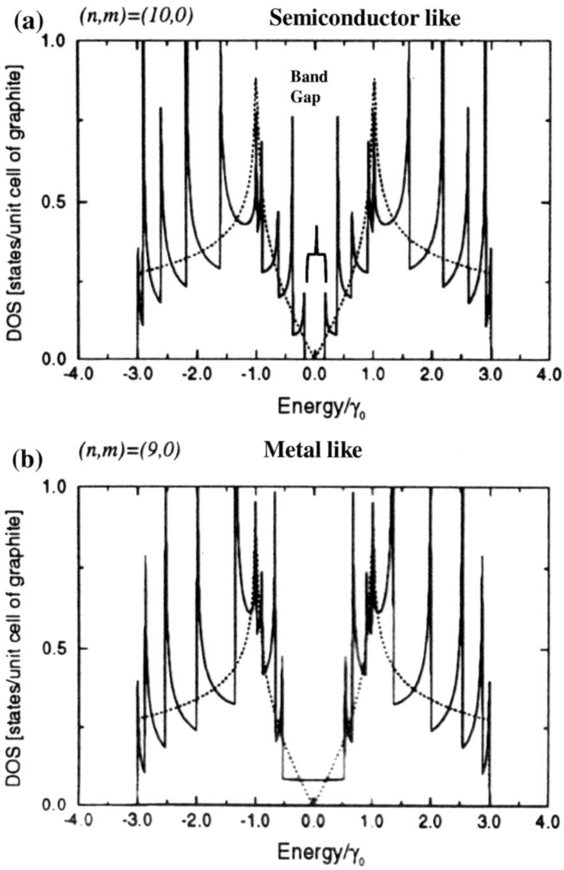

Fig. 1.18 Density of states (DOS) of CNT. **a** Semiconductor CNT. **b** Metallic CNT. The density of states of grapheme is also shown as *dotted line* for the reference [14]

been understood, but the electron state of CNT is thought to be changed by touching the CNT on the substrate.

Figure 1.20 shows luminescence from single-layer CNT. Because excitation light is irradiated to the large area of the grass-like CNT, the emission spectrum from CNT with different diameter is observed as shown in Fig. 1.20 [19]. Figure 1.21 shows inclination characteristic of luminescence of CNT. The energy of the excitation light is with 0.5 mW power at 1.481 eV in the spot size of 2 μm [19]. (a) Shows emission spectrum strength when the angle from the direction of length of CNT is changed. The inclination characteristic of luminescence is clearly shown. (b) Shows an angle dependence of excitation light on the luminescence intensity. A strong inclination characteristic is observed. The solid line is the one having drawn by the function of $\cos 2\theta = 0.7$.

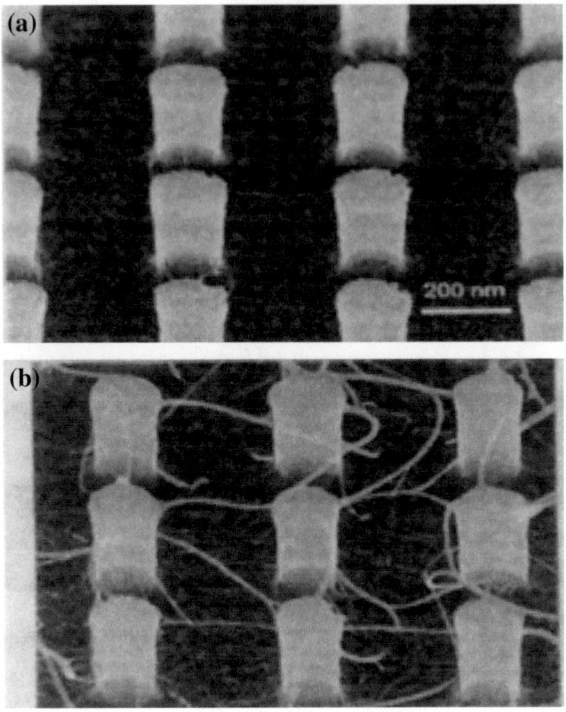

Fig. 1.19 SEM image of grown single wall CNT (**a**) and bundled CNT (**b**). (1.18) [18]

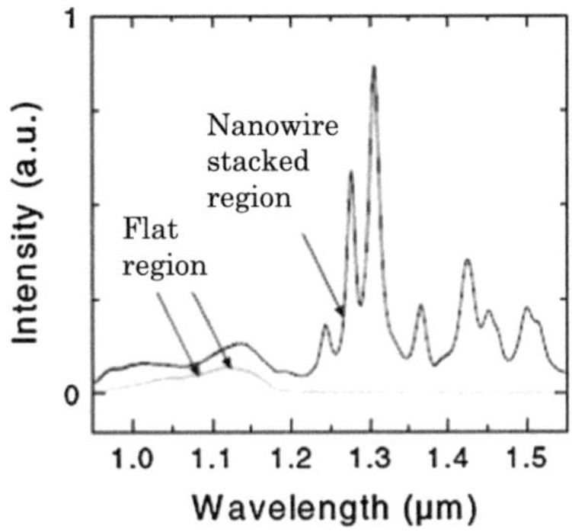

Fig. 1.20 Emission spectrum from single-layer carbon nanotubes. In this case large area of CNT mat was irradiated by excitation light of He-Ne laser of 1.7 mW with spot size of 2 μm. So, luminescence from CNT with different diameter was observed as different peaks. When CNT is put on the substrate, no emission was observed [17]

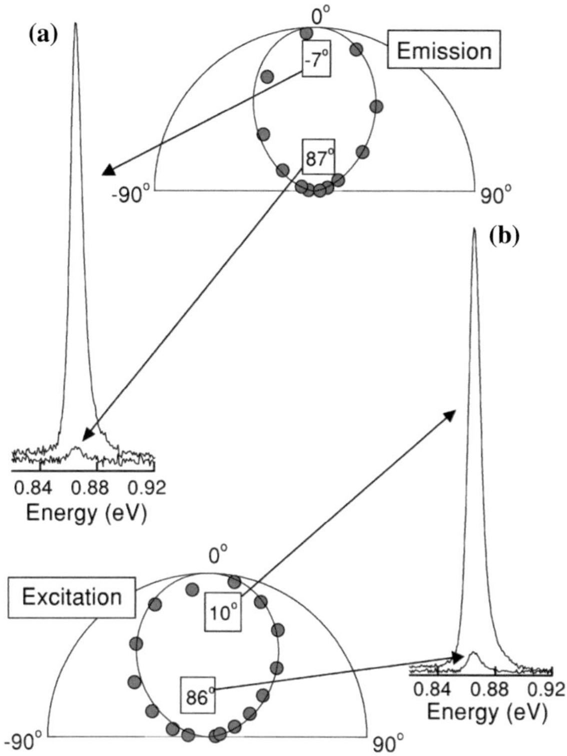

Fig. 1.21 Poralization and angular dependence of emission light from CNT. Energy of excitation is 1.481 eV, 0.5 mW with spot size of 2 μm [19]

The technology that the nanopillar of semiconductor is formed by a selective growth method or a surfactant method is established recently and the optical properties have come to be examined though the details are not described here [20, 21]. A typical application of the quantum wire is a quantum-wire laser. Let me explain the quantum-wire laser later.

1.3.4 Optical Property of Quantum Dot

Since quantum dot has density of state of delta function, it is expected in early stage of study of the luminescence from quantum dot gas like sharp luminescence to be observed. However, for a long time, such spectrum had not been observed and it was a mystery. It becomes recently clear that very narrow spectrum of a line width of 50 μeV can be observed [21] by using a spectroscope with very high resolution under very weak light excitation. Figure 1.22 shows the one example.

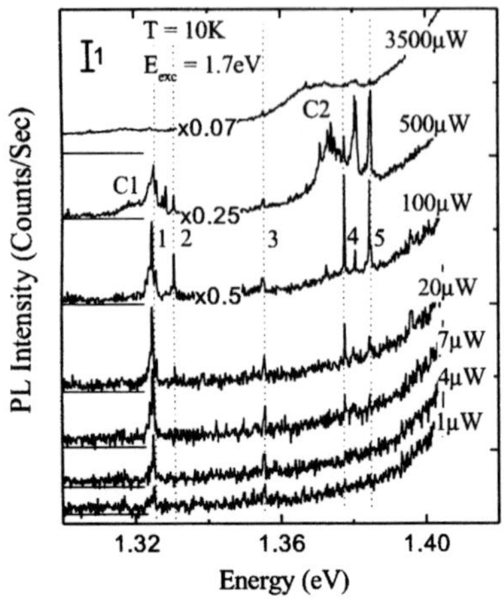

Fig. 1.22 Excitation intensity dependence of the luminescence from quantum dot

The experimental results of luminescence from quantum dot are well explained the transition calculated for the quantum dot shown in Fig. 1.23 [22]. Moreover, it has been understood that luminescence from the single quantum dot blinks in a certain condition as shown in Fig. 1.24 (b), (c) [23, 24]. It has been understood that similar phenomena has happened in various materials [24, 25].

Moreover, it is understood that this blinking has a nature of telegram noise as shown in Fig. 1.25. Figure 1.25 also shows the strength change of luminescence. The interval of the blinking decreases at time and the earlier the time is and stronger the excitation strength is, the faster the switching rate is. It is thought that there is a kind of trap in the quantum dot and the electron may go in and out into the trap. The electron trapped in the quantum dot changes the electric field in the dot and it causes a modification of the electron and hole wave function spatially in succession of large change of luminescence. Therefore, it is thought that trapping and de-trapping into the defect may be important for the blinking phenomena. Similar blinking luminescence is observed as for luminescence from carbon nanotubes recently.

1.4 Application of Quantum Structures Toward Optical Devices

1.4.1 Cascade Laser

Let's show examples of recent application of a super-lattice here. It is proposed a semiconductor laser that leaves a concept quite different from a conventional semiconductor laser principle in recent, that is, quantum-cascade laser (QCL). It is produced by applying many kinds of super-lattices to a semiconductor laser [26–30]. QCL tries to obtain the laser using the transition between sub-bands in a super lat-

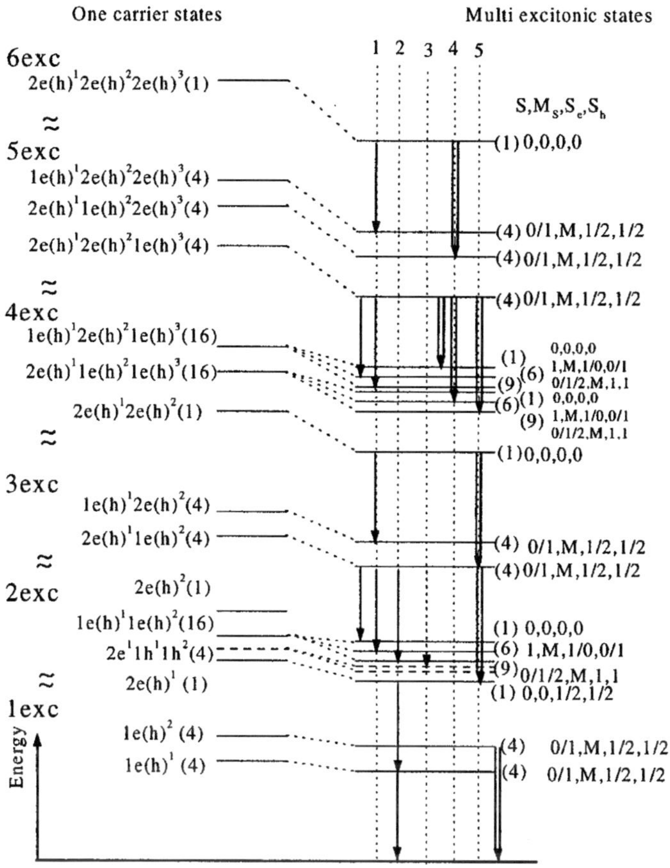

Fig. 1.23 Calculated transition expected in quantum dot [22]

tice. The research on THz is comparatively advanced from the wavelength of several microns in the area of the long wavelength as for this now. Especially, it is expected that a high performance is put out in the THz area and is interesting as the source of light of various applications of the THz area. Because the laser in the THz area can penetrate the material that cannot usually be penetrated in the visible region. So, various detection, applications to the medical treatment, and other applications of variety are thought. The principle of QCL is shown in the Fig. 1.26 [29]. This figure shows an oscillation principle of quantum cascade laser [28].

QCL uses the inter sub-band transition in the multi-quantum well as shown in the figure. It became possible to oscillate the laser by designing the inter sub-band of multi-quantum well though the inter sub-band transition of the quantum well was thought that it was difficult to make the population inversion to cause the laser oscillation because it was considered to be very fast usually. The above figure shows the typical example of QCL. The width of the quantum well is narrowly designed in

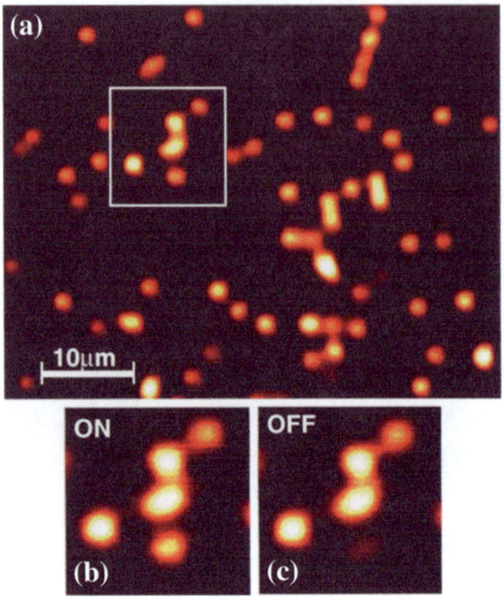

Fig. 1.24 Brinking phenomena of luminescent light from quantum dots. In two figures of *bottom* the clear spontaneous light turn on and off is observed [24]

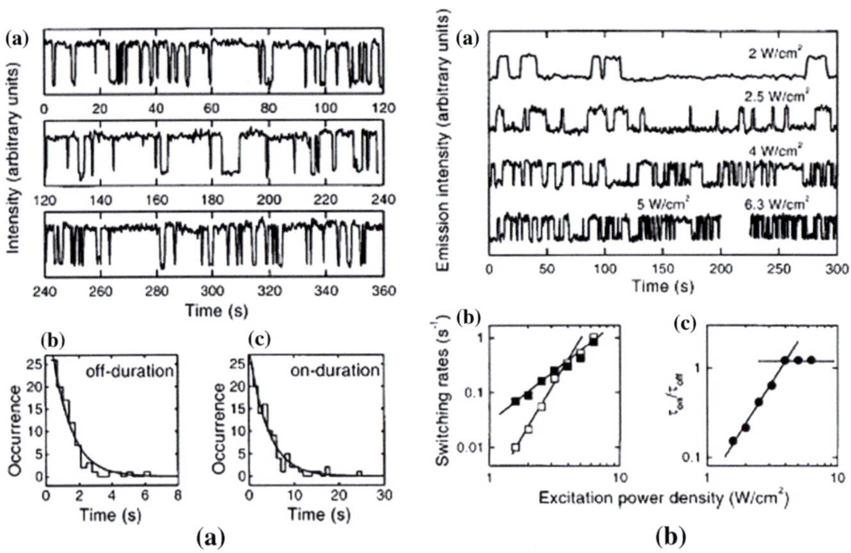

Fig. 1.25 Time sequence of the blinking luminescent light form a quantum dot and occurrence probability of the phenomena as a function of time. In this figure the excitation power density dependence of switching rates and on/off ratio are also shown [24]

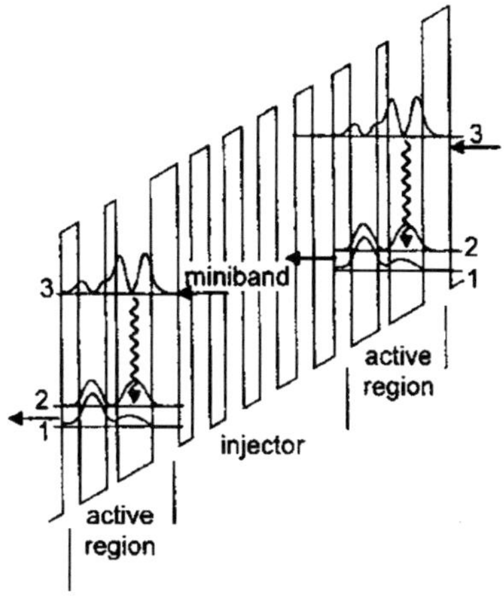

Fig. 1.26 Schematic diagram of oscillation principle of quantum cascade laser [28]

the injection area, and the first quantum state exists comparatively in a high energy level and the mini-band is formed with the interaction between quantum wells. The active regions where laser oscillation happens are connected in both sides of the injector region as shown in Fig. 1.26 to get an efficient laser oscillation. The career in the injection region is injected into the active region. That is, the electron is propagated the mini-band formed in the injection region, and reaches the following active region. The laser oscillation can be efficiently obtained by actually making multisteps of this combination of injection region and active region to re-inject the carrier from one step to another step. This multi injection of carrier into the active region makes possible the laser oscillation. It takes as an example of the case of InGaAs/AlInAs that typically adjusts the lattice to the InP substrate. When thinking about $In_{0.53}Ga_{0.47}As/Al_{0.48}In_{0.52}As$ super-lattice, the energy difference between barrier and well is 520 meV. If we consider the quantum well thickness of the active layer of 6.0, 4.7 nm, and barrier layer of 1.6 nm, the energy difference between level 2 and 3 is 207 meV (correspond by six microns in wavelength) and level 2 and 1 corresponds to 37 meV. From the Frolich model collected electrons in level 3 relaxes to the level 2 and level 1 with a relaxation time of $\tau_{32} = 2.2$ ps and $\tau_{31} = 2.1$ ps, respectively. This relaxation times give us the relaxation time of level 3 of $\tau_3 = 1.1$ ps$(\tau_3 = 1/\tau 31 + \tau_{32})$. The energy between level 2 and 1 is designed to be the same as the LO phonon energy and the relaxation time between level 2 and 1 $\tau_{21}(= \tau_2)$ is 0.3 ps and is quite short. So, the population inversion between level 3 and 2 can be easily formed. If the electron in injection region injects into the active region quickly and the electron relaxed to the ground state of active region and diminished to the next injection region by the tunneling between two regions, the CW operation is possible. Typically QCL with 20 to 30 active regions is designed and recently the QCL with more than 100 active regions has tried to be fabricated. Figure 1.27 is a cross sectional TEM image of a super-lattice of QCL [28].

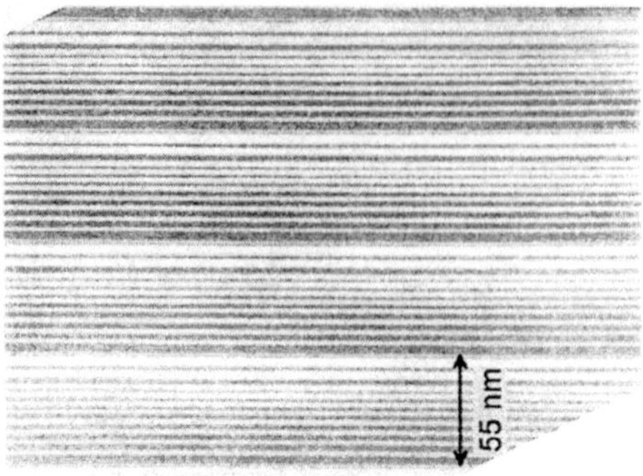

Fig. 1.27 Cross sectional view of super lattice used in quantum cascade laser [28]

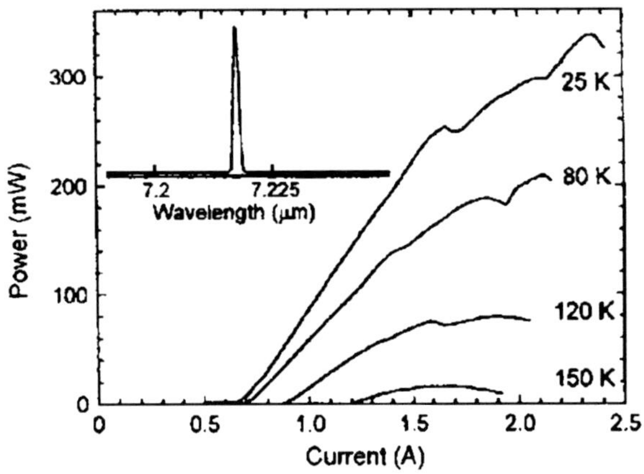

Fig. 1.28 Injection current dependence of output power of quantum cascade laser as a function of operating temperature. Inset shows oscillation spectrum of quantum cascade laser [28]

Quite recently the several modifications for designing the multi- super lattices have been tried. One of the examples is chirped structure of injection region and RT operation of QCL has been observed. Figure 1.28 shows the CW oscillation spectrum of QCL and an example of the temperature dependence. This device uses chirped super-lattice with RIDGID type structure. A lot of papers has gone out for the cascade laser. The example of the references has been previously shown [26–30].

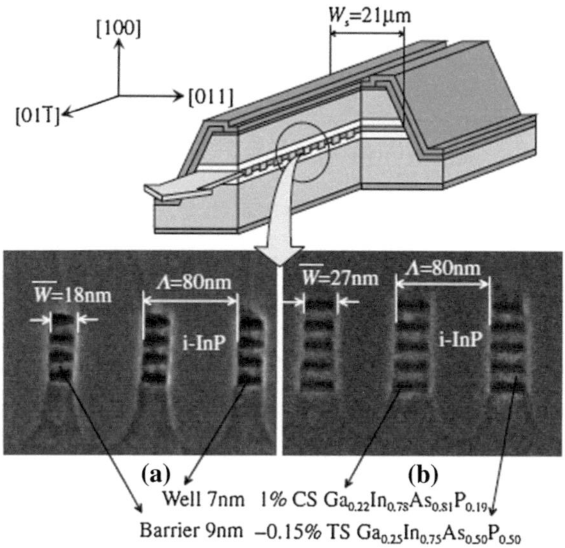

Fig. 1.29 Typical structure of quantum wire laser and cross sectional SEM view of (**a**) quantum wires 18 and (**b**) quantum wire 27 used in this device structure [33]

1.4.2 Quantum Wire Laser

The quantum-wire laser is analyzed by Asada et. al. [31] in details. Here, experimental results are mainly introduced. Recently, the research of the quantum wire laser has been making advances and the laser oscillation has become possible [32–35]. Let's explain here a method [32] of producing quantum wire using electron beam lithography [33], though there are a couple of method to make the quantum wire, that is, the spontaneous formation method of the quantum wire, a method using wet chemical etching [32] etc. The structure of a typical quantum-wire laser is shown in Fig. 1.29 [33].

The quantum wire is produced by using four composites materials of GaInAsP. In the figure the quantum wire of 18 layers and 27 layers stacking case are shown. In Fig. 1.30 lasing intensity as a function of injection current is described.

Injection current dependency [33] of laser oscillation strength of quantum wire laser as a function of numbers of stacked quantum wires are described in Fig. 1.30. Ws is width of the stripe of the laser. When the quantum wires are increased up to 70, it is understood that the threshold is below the half compared with a quantum well laser.

Filling rate versus normalized threshold current density in quantum wire laser is described [33] in Fig. 1.31 as a function of surface recombination velocity of S. The filling factor ρ of the laser are given by W/Λ. As for the recombination at surfaces, the threshold of the quantum-wire laser increases to more than the cases of the quantum well for $S = 300\,\mathrm{cm/s}$ that comes out along with the processing making of the

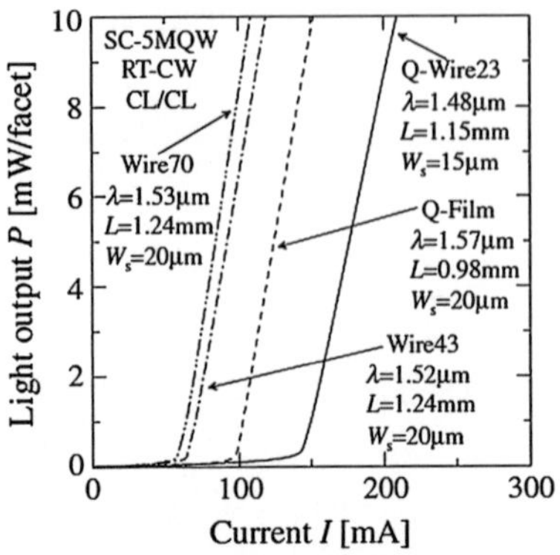

Fig. 1.30 Injection current dependence of oscillation intensity of quantum wire laser with wire number of 23, 43, 70 observed RT-CW condition. The case of quantum well is also described as a reference. Ws is the width of stripe of quantum wire [33]

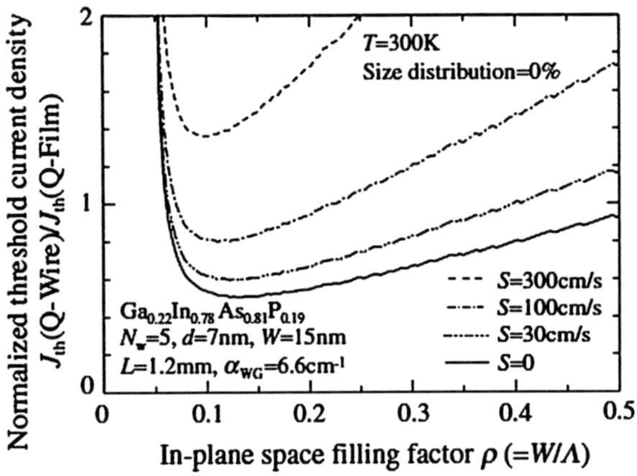

Fig. 1.31 In-plane space filling factor (ρ) dependence of the relative threshold current density. S is the surface recombination velocit [33]

quantum wire. It is at least necessary that the value of the surface recombination rate is 30 cm/s or less. Moreover, it is understood that the best filling rate exists.

1.4.3 Quantum Dot Laser

The temperature dependence of the threshold of the laser with the quantized structure was predicted from early time by Arakawa [36, 37]. It comes to calculate this of the quantum well, quantum wire, and quantum dot in case of GaAs/AlGaAs etc. as shown in the Fig. 1.32. The threshold of the laser using the quantum dot doesn't

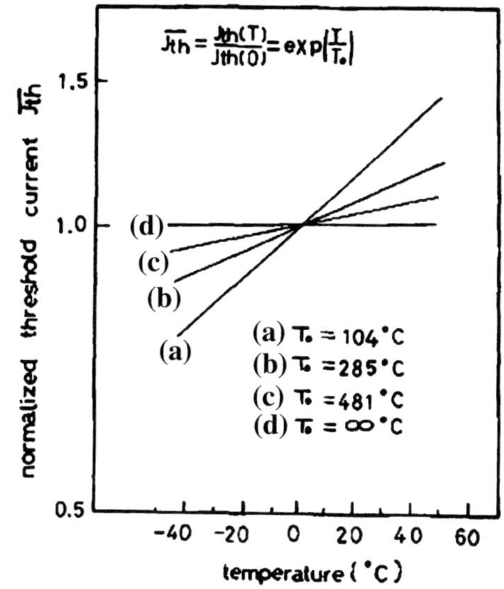

Fig. 1.32 Temperature dependence of the threshold of quantum dots laser. **a** Normal laser, **b** quantum well laser, **c** quantum wire laser, **d** quantum dot laser [36]

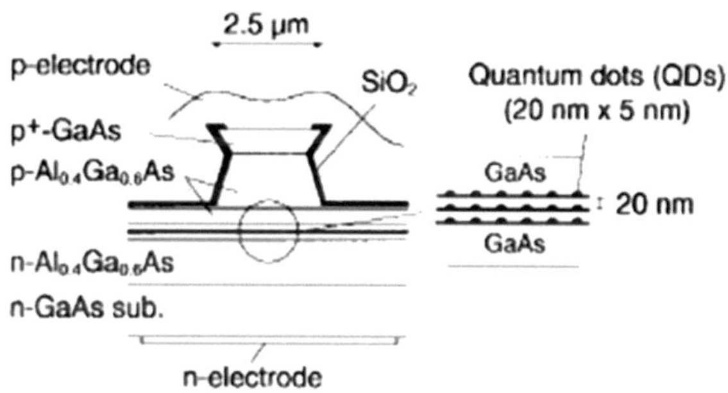

Fig. 1.33 Schematic cross section of ridge-mesa quantum dot laser [39]

have the temperature dependence ((d) of Fig. 1.32), and the temperature dependence becomes larger when the quantum wire (c) and quantum well (b) is used. (a) is a usual semiconductor laser in this figure. The electron is, however, here only in the ground state of the sub band of the quantum structure and electron in excited state is not considered.

The CW semiconductor laser with a little temperature dependence has been realized recently by the development of the technology of formation of quantum dots [38–42].

One example of the structure of the quantum dot laser with the ridge mesa structure is shown in Fig. 1.33 (1.39). The InGaAs quantum dot naturally formed is accumu-

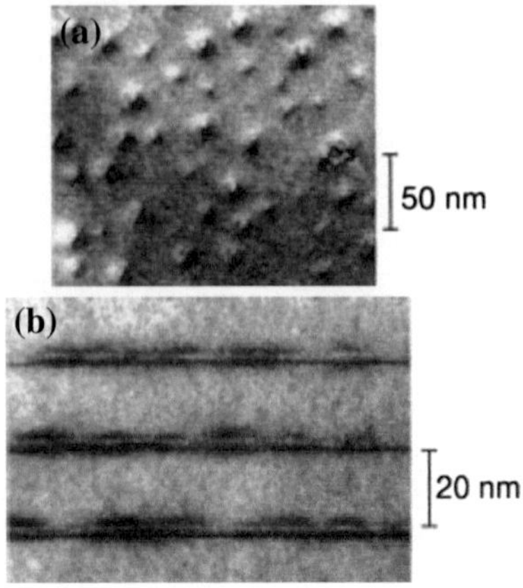

Fig. 1.34 Transmission electron microscope photographs of a triple-stacked quantum-dot structure. **a** A plan view. **b** A cross-sectional view [39]

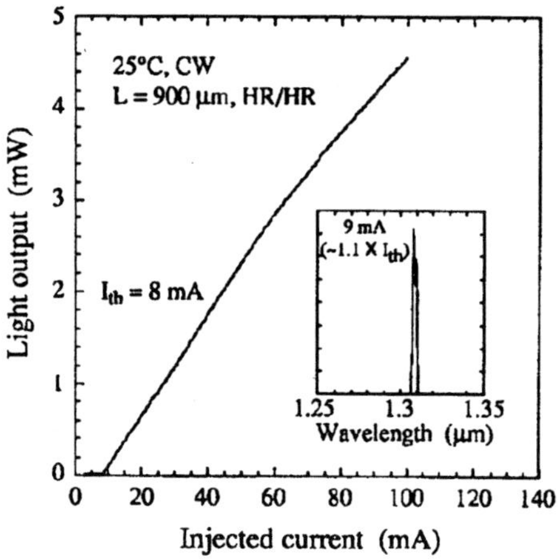

Fig. 1.35 CW light output versus injected current characteristics of laser with a 900 μm cavity length and high-reflectivity-coated facets. Threshold current (I_{th}) was 8 mA. Lasing spectra at 9 mA ($\sim 1.1 \times I_{th}$) is shown in the inset, indicating a lasing wavelength of 1.31 μm [39]

lated to the active layer by three layers at intervals of 20 nm. The low thresholding by an enough current narrow channel stricture is achieved by taking the ridge structure for the GaAs quantum dots laser. Figure 1.34 shows quantum-dot structure [39] used in this device. It is understood for the quantum dot naturally formed in the SK mode shown in Fig. 1.34 to be able to extend to three layers with almost self-aligned structure. One example and the temperature property of the oscillation characteristic are shown in Figs. 1.35 and 1.36, respectively.

It is understood that a single mode oscillation at the level with low injection current was observed, that is, the area of $1.1 \times I_{th}$. However, when the injection current is increased, the oscillation from the sublevel of the quantum dot comes to be observed. This is because the oscillation from a sub-band in the quantum dots and/or quantum well appears as soon as the injection current increases because the number of electron which can be occupied in the dots is small and the carriers infected into the

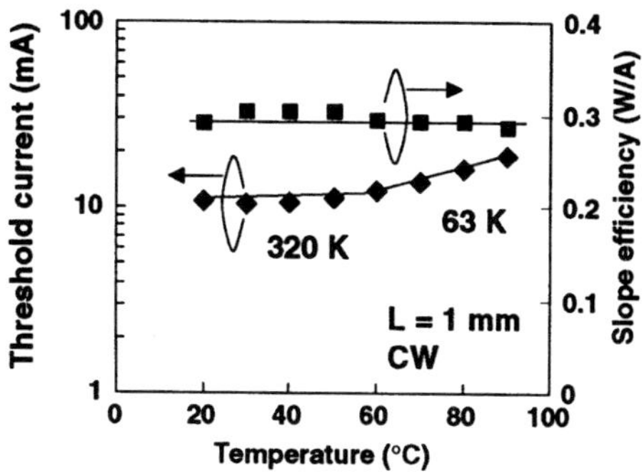

Fig. 1.36 Temperature dependence of threshold current and slope efficiency [42]

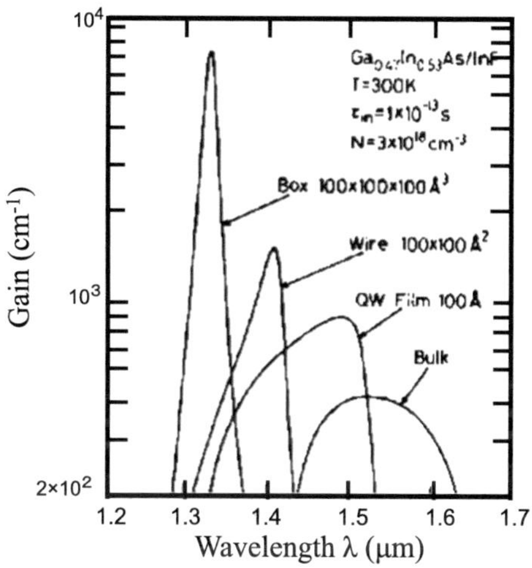

Fig. 1.37 Gain of laser oscillation for quantum dot laser, quantum wire laser and quantum well laser [49]

dots are easily overflow to the outside of the dots. Adetailed theoretical analysis on these oscillations is performed politcly in Sugawara's et al. [40]. Moreover, threshold current of the quantum dot and temperature dependence of the slope efficiency [42] are shown in Fig. 1.36.

A low temperature dependence is observed from 20 to 60 C. Moreover, the slope efficiency has not changed over 90° from 20 C. The laser is insensitive against the temperature and is expected the steady operation in all the temperature environments is possible. This characteristic is an important for the optical information processing and the optical communication in the future. The person who wants to learn more detailed information for the quantum dot laser must refer to document [41–48].

When a similar laser is made by routine ways, characteristic temperature T_0 is about 70 K and slope efficiency is -1 dB. Comparing to the conventional laser the performance of the quantum dots laser has improved greatly.

Fig. 1.38 Theoretical calculation of maximum gain for quantum well, quantum wire, quantum dot and balk laser for (**a**) GaInAs/InP, (**b**) GaAs/GaAlAs

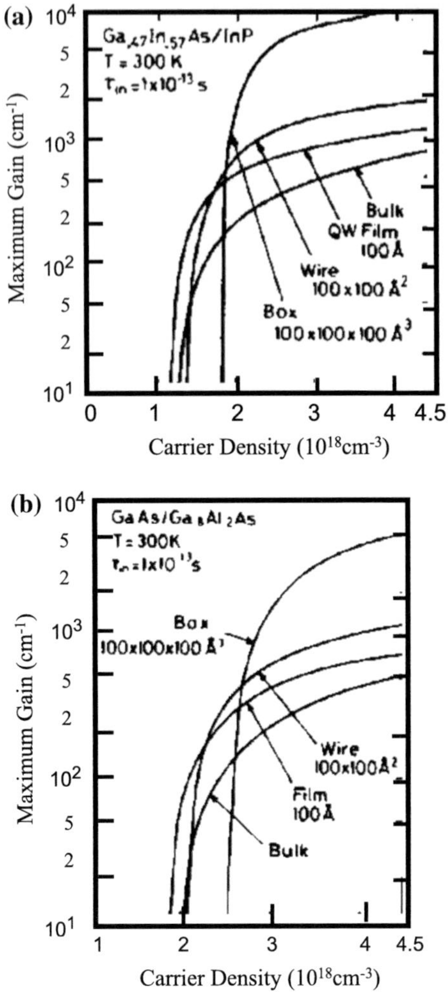

1.4.4 Gain of Quantum Structure Laser

The gain of the quantum dot laser was calculated by Asada et al. [52], the results are shown in Fig. 1.37. In Fig. 1.37 the calculated gain of the quantum dot of 10 × 10 × 10 nm, the quantum wire of 10 × 10 nm and the quantum well of 10 nm using materials of $Ga_{0.47}In_{0.53}As/InP$, [49] are shown. It is understood the gain for the quantum dot is very high compared with the quantum wire and the quantum well. In Fig. 1.38 the calculated results of the maximum gain for injection current are described. The details of the calculation are written in the references [31, 51, 52].

The numerical result of the maximal gain to the injection current is shown in Fig. 1.38 for GaIn As/InP (a) and (b) GaAs/GaAlAs, for the quantum wires, the quantum dots, and bulks. The related references should be followed in [31, 50–52].

References

1. A.G. Milnes, D.L. Feucht, *Semiconductor Heterostructures*, (Pergamon Press, Oxford, 1974)
2. R. Dingle, W. Wiegmann, C.H. Henry, Phys. Rev. Lett. **33**, 827 (1974)
3. G. Bastard, E.E. Mendz, L.L. Chang, L. Esaki, Phys. Rev. B **26**, 1974 (1982)
4. T. Ishibashi, Y. Suzuki, H. Okamoto, Jpn. J. Appl. Phys. **20**, 632 (1981)
5. M. Shinada, S. Sugano, J. Phys. Soc. Jpn **21**, 1936 (1966)
6. H. Iwamura, H. Kobayashi, H. Okamoto, Jpn. J. Appl. Phys. **23**, 795 (1984)
7. T. Miyazawa, S. Tarucha, Y. Ohmori, Y. Suzuki, H. Okamoto, Jpn. J. Appl. Phys. **25**, 200 (1986)
8. Y. Kawaguchi, H. Asahi, Appl. Phys. Lett. **50**, 1243 (1987)
9. A.S. Barker Jr., A.J. Sievers, Rev. Mod. Phys. **47**, Suppl.2 (1975)
10. M. Razeghi, J. Nagle, P. Maurel, F. Omnes, J.P. Pocholled, Appl. Phys. Lett. **49**, 1110 (1986)
11. J.S. Weiner, D.S. Chemia, D.A.B. Miller, T.H. Hood, D. Sivco, A.Y. Cho, Appl. Phy. Lett. **46**, 619 (1985)
12. O. Akimoto, H. Hasegawa, J. Phys. Soc. Jpn **22**, 181 (1967)
13. T. Ando, T. Nakanishi, *Carbon Nanotube and Quantum Effects* (Iwanami, Tokyo, 2007), pp. 18
14. R. Saito, G. Dresselhaus, M.S. Dresselhsus, *Physical Properties of Carbon Nanotubes* (Imperial College Press, London, 2003), pp. 68
15. M.J.O. Connell, S.M. Bachilo, C.B. Huffman, V.C. Moore, M.S. Strano, E.H. Haroz, K.L. Rialon, P.J. Boul, W.H. Noon, C. Kittrell, J. Ma, R.H. Hauge, R.B. Weisman, R.E. Smalley, Science **297**, 593(2002)
16. S.M. Bachilo, M.S. Strano, C. Kittrell, R.H. Hauge, R.E. Smalley, R.B. Weisman, Science **298**, 2361 (2002)
17. J. Lefebvre, M.M. Fraser, Y. Homma, P. Finnie, Phys. Rev. Lett. **90**, 217401 (2003)
18. Y. Homma J. Lefebre, P. Finnie, Phys. Solid State **39**, 170(2004) (in Japanese)
19. J. Lefebvre, M.M. Fraser, P. Finnie, Y. Homma, Phys. Rev. B **69**, 075403–1 (2007)
20. H. shiozawa, H. Ishii, H. Ktataura, H. Yoshioka, H. Kihara, Y. Takayama, T. Miyahara, S. Suzuki, Y. Achiba, T. Kodama, M. Nakatake, T. Narimura, M. Higashiguchi, K. Shimada, H. Namatame, M. Taniguchi, Physica B **351**, 259(2004)
21. A.G. Rozhin, Y. Sakakibara, M. Tokumoto, H. Kataura, Y. Achiba, Thin Solid Films **368**, 464 (2004)
22. E. Dekel, D. Gershoni, E. Ehrenfreund, D. Spektor, J.M. Garcia, P.M. Petroff, Phys. Rev. Lett. **80**, 4991(1998)
23. M. Nirmal, B.O. Dabbousi, M.G. Bawendi, J.J. Macklin, J.K. Trautman, T.D. Harris, L.E. Brus, Nature **383**, 802 (1996)
24. M.E. Pistol. P. Castrillo, D. Hessman, J.A. Prieto, L. Samuelson, Phys. Rev. B **59**, 10725(1999)
25. M. Sugisaki, H.W. Ren, K. Nishi, Y. Masumoto, Phys. Rev. Lett. **86**, 4883 (2001)
26. R. Kazarinov, R.A. Suris, Fiz. Tekh. Poluprovodn. **5**, 797(1971)(Sov. Phys. Semicond. **5**, 707, (1971))
27. J. Faist, F. Capasso, D.L. Sivco, C. Sirtori, A.L. Hutchinson, A.Y. Cho, Science **264**, 553 (1994)
28. C. Gmachl, F. Capasso, L. Sivco, A.Y. Cho, Rep. Prog. Phys. **64**, 1533 (2001)
29. J. Faist, F. Capasso, D.L. Sivco, A.L. Hutchinson, S.N.G. Chu, A.Y. Cho, Appl. Phys. Lett. **72**, 680 (1998)
30. R. Colombelli, A. Straub, F. Capasso, C. Gmachl, M.I. Blakey, A.M. Sergent, S.N.G. Chu, K.W. West, L.N. Pfeiffer, J. Appl. Phys. **91**, 3526 (2002)

31. M. Asada, Y. Miyamoto, Y. Suematsu, Jpn. J. Appl. Phys. **24**, 95 (1985)
32. K.C. Shin, S. Arai, Y. Nagashima, K. Kudo, S. Tamura, IEEE Photonics Technol. Lett. **7**, 345 (1995)
33. H. Yagi, T. Sano, K. Ohira, D. Plumwongrot, T. Maruyama, A. Haoue, S. Tamura, S. Arai, Jpn. J. Appl. Phys. **43**, 3401 (2004)
34. Y. Ohno, M. Higashiwaki, S. Shimomura, S. Hiyamizu, S. Ikawa, J. Vac. Sci. Technol B **18**, 1672 (2000)
35. A. Haque, T. Maruyama, H. Yagi, T. Sano, P. Dhanorm, S. Arai, IEEE J. Quantum Electronics **40**, 1344 (2004)
36. Y. Arakawa, H. Sasaki, Appl. Phys. Lett. **40**, 393 (1982)
37. Y. Arakawa, H. Sakaki, M. Nishioka, Oyo Buturi **52**, 852(1983)(in Japanese)
38. K. Mukai, Y. Nakata, K. Otsubo, M. Sugawara, N. Yoshikawa, H. Isikawa, IEEE, Photonics Technol. Lett. **11**, 1205(1999)
39. H. Syohi, Y. Nakata, K. Mukai, Y. Sugiyama, M. Sugawara, N. yoshikawa, H. Isikawa, IEEE J. Sel. Top. Quantum Electronics, **3**, 188(1997)
40. M. Sugawara, H. Ebe, N. Hatori, M. Ishida, Y. Arakawa, T. Akiyama, K. Otsubo, Y. Nakata, Phys. Rev. B **69**, 235332 (2004)
41. M. Sugawara, N. Hatori, H. Ebe, M. Ishida, Y. Arakawa, T. Akiyama, K. Otsubo, Y. Nakata, J. Appl. Phys. **97**, 043523 (2005)
42. K. Otsubo, N. Hatori, M. Ishida, S. Okumura, T. Akiyama, Y. Nakata, H. Ebe, M. Sugawara, Y. Arakawa, Jpn. J. Appl. Phys. **43**, 1124 (2004)
43. M. Ishida, N. Hatori, T. Akiyama, K. Otsubo, Y. Nakata, H. Ebe, M. Sugawara, Y. Arakawa, Appl. Phys. Lett. **85**, 4145 (2004)
44. T. Akiyama, M. Ekawa, M. Sugawara, K. Kawaguchi, H. Sudo, A. Kuramata, H. Ebe, Y. Arakawa, IEEE Photonics Technol. Lett. **17**, 1614 (2005)
45. K. Mukai, N. Ohtsuka, M. Sugawara, S. Yamazaki, J. Appl. Phys. **33**, 1710 (1994)
46. M. Sugawara, K. Mukai, Y. Nakata, H. Isikawa, Phys. Rev. B **61**, 7595 (2000)
47. Y. Arakawa, M. Sugawara, Proc. of SPIE **5722**, 45 (2005)
48. V. Tokranov, M. Yakimov, A. Katsnelson, M. Lamberti, S. Oktyabsky, Appl. Phys. Lett. **83**, 833 (2003)
49. M. Asada, Y. Miyamoto, Y. Suemastu, IEEE J. Quantum Electonic **QE22**, 1915(1986)
50. M. Asada, A. Kameyama, Y. Suematsu, IEEE J. Quantum Electonic **QE20**, 745(1984)
51. M. Asada, Y. Suematsu, IEEE J. Quantum Electonic **QE21**, 434(1985)
52. M. Yamada, Y. Suematsu, J. Appl. Phys. **52**, 2653 (1981)

Chapter 2
Photonic Crystals: Manipulating Light with Periodic Structures

Shin-ichiro Inoue

Abstract The new physical phenomena being discovered in photonic crystals during the last two decades have revealed that photonic crystals enable light manipulation, strong light confinement, slow light propagation, and enhanced light-matter interaction, which are unique and cannot be achieved in conventional materials. Due to these properties and their potential technological applications, photonic crystals have become the subject of intense research. In this chapter, starting with a brief review of the photonic crystal concept, we describe fundamental features, fabrication technologies, experimental advances, and potential impact of photonic crystals. The progress achieved in photonic crystal research is closely connected to the evolution of device design concepts through a deeper understanding of nature of photonic bands. We describe detailed methods for the numerical and experimental determination of the photonic band structure that are useful and essential for analyzing photonic crystals as well as for understanding and engineering their unprecedented optical properties. This chapter closes with a discussion of the exciting possibilities of a photonic crystal in nonlinear light-matter interactions and their device applications.

2.1 The Concept of Photonic Crystal

Photonic crystals are materials composed of dielectric structures with periodicity in one, two or three directions on the optical wavelength scale. The ability of a photonic crystal to control the light dispersion relation (i.e., photonic band structure) with a high degree of freedom is an issue of scientific and practical importance. Tailoring of the band dispersion in photonic crystal systems can give rise to anomalous dispersion characteristics including photonic band gap and extremely slow

---

S. Inoue (✉)
Advanced ICT Research Institute, National Institute of Information and Communications Technology (NICT), 588-2 Iwaoka, Nishi-ku, Kobe 651-2492, Japan
e-mail: s_inoue@nict.go.jp

Y. Aoyagi and K. Kajikawa (eds.), *Optical Properties of Advanced Materials*,
Springer Series in Materials Science 168, DOI: 10.1007/978-3-642-33527-3_2,

group velocities (a phenomenon known as "slow light") that cannot achieved in homogeneous materials and conventional waveguides. These characteristics have opened up unprecedented and exciting possibilities in a wide range of photonics with multiple applications in photonic information and communication, laser technologies, optoelectronics, integrated, and nonlinear optics.

Various forms of periodic dielectric structure have been studied scientifically for more than a century. The first concept of one-dimensional photonic crystals is formation of a stop band in multilayer dielectric films (such as the Bragg reflection mirror), which has been proposed by Lord Rayleigh in 1887 [1, 2]. Beyond one-dimensional concept, C. G. Darwin (grandson of Charles Darwin) identified dynamical theory of X-ray diffraction with the specific features of the perfect reflection for a certain range of angles (i.e., narrow stop bands for X-ray waves in three-dimensional crystals) in 1914 [3]. In optics, the theoretical model of three-dimensional dielectric lattice and energy band of photons have been discussed by Ohtaka in 1979 [4]. The name photonic crystal and the exciting growth of photonic crystal research began with the clear concept and work of E. Yablonovitch (Bellcore, presently at UCLA) and his group in the late 1980s [5, 6]. Initially, photonic crystal was introduced in the context of controlling spontaneous emission of materials embedded within periodic media. And S. John (Princeton Univ. presently at Toronto Univ.) reported at almost the same time that disorder-induced strong light localization occurs in photonic crystal structure with high index contrast [7]. By these discoveries, it was proved that photonic crystals are artificial crystal for photons, and in many respects the behavior of photons in photonic crystals is similar to the behavior of electrons in a periodic Coulomb potential of a solid semiconductor crystal [8]. Therefore, the concept of photonic crystals has been developed using an analogy between the propagations of electrons in solid semiconductor circuits and electromagnetic waves in photonic crystal circuits.

Since a crystal for photons is a periodic arrangement of potential distribution (refractive index contrast of dielectric arrays), this lattice introduces the photonic band structure of the crystal due to Bragg-like diffraction of electromagnetic wave. From general symmetry principles, the electromagnetic modes of a photonic crystal with periodicity can be written as Bloch states in Maxwell's equations. Here, the functions of the mode profiles are determined by the eigenvalue problem restricted to a single unit cell of the photonic crystal because of the boundary condition. These Bloch boundary conditions over a unit cell of a photonic crystal give the k vector dependence of the eigen-mode, and the spectral information can be obtained by a Fourier transform of the fields. The information contained in these functions is called the photonic band structure of the photonic crystal. Figure 2.1 shows the dispersion curve for an electromagnetic wave in a homogeneous medium and an example of the photonic band dispersion in a two-dimensional photonic crystal. The calculation techniques that were used to generate the theoretical photonic band structure and its details are outlined in Sect. 2.3. If the lattice potential is strong enough, the photonic band gap is produced into the photonic band structure. In this photonic band gap frequency region, the propagation of electromagnetic wave is forbidden in certain or all directions. K. M. Ho, C. M. Soukoulis et al. (Iowa State Univ.) found for the

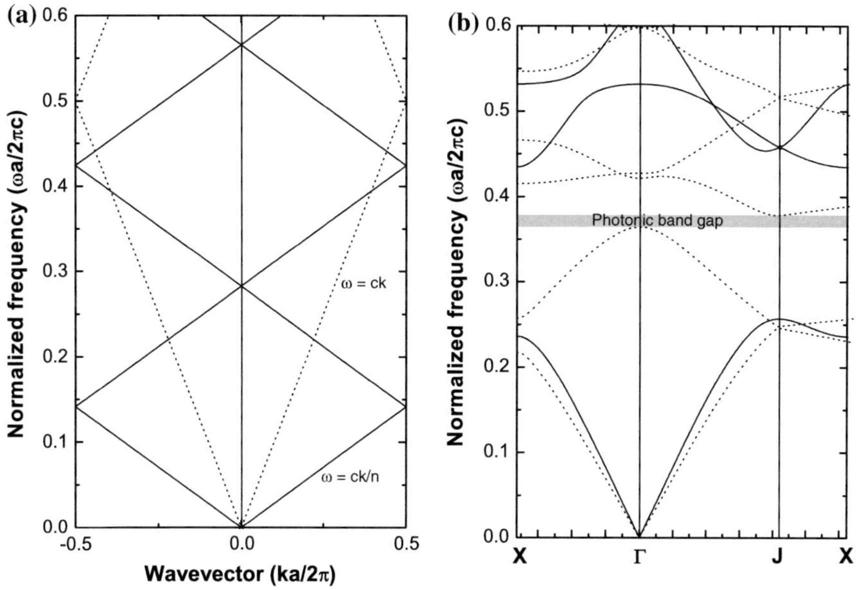

Fig. 2.1 Comparison of dispersion relation in a homogeneous medium to photonic band structure in a 2D photonic crystal. **a** Dispersion line of a uniform material with dielectric constant $\epsilon = 12.5$ (*solid line*) and *light line* in a vacuum (*dotted line*). **b** Photonic band structure of the 2D photonic crystal with the triangular lattice ($\epsilon = 12.5$) for TE (*solid line*) and TM polarizations (*dotted line*). It has the photonic band gap for both TE and TM modes (the complete photonic band gap)

first time that the three-dimensional photonic crystal in a diamond lattice structure generated a complete photonic band gap [9].

The photonic band structure and band gaps thus can be firmly understood in a classical framework based on Maxwell's equations without a need to resort to quantum mechanics. However, this does not prevent a very exciting and often unexpected variety of optical phenomena from occurring, due to the anomalous dispersion characteristics and its complicated law in a two- or three-dimensional photonic crystal structure with high-refractive index contrast.

2.2 Technology to Fabricate Photonic Crystal

A broad range of biological periodic dielectric structures with interference-based structural colors can be found in nature. For example, multilayer structures are known to produce vivid iridescent coloration in certain butterflies [10–12]. The spectacular metallic blue and green colors are present in of the wing scale of the tropical Morpho butterfly, which is mainly from the Bragg reflection of the one-dimensional photonic structure. A natural two-dimensional periodic lattice can be found in a sea mouse [13]. Strong directionally dependent Bragg scattering produces intense iridescence from a sea mouse hair including internal two-dimensional periodicity that may be called a naturally occurring photonic crystal fiber. Recently, a natural three-dimensional photonic structure with a diamond-based crystal lattice operating at visible wavelengths was discovered in scales of a Brazilian beetle [14]. Periodic micro- and nanostructures and their optical functions in nature have been optimized as the continuation of biological evolution over several hundreds of millions of years.

A universal feature of biological photonic structures is a low-refractive index contrast. Periodic micro- and nanostructures that nature provides can be used as templates for fabricating a wide range of photonic related structures [15–17]. Inorganic structures with high-index contrast replicated from biological templates can combine the merits offered by both the material and biological structures. Naturally occurring periodic dielectric structures can be also found in inorganic materials, such as opals. Natural opal is composed of silica spheres in a hexagonal or cubic close-packed lattice on the scale of optical wavelength, however, which does not exhibit multi-directional photonic band gap. Artificial opals can be fabricated by self-assembly of colloidal polymer spheres [18–20]. Template colloidal techniques using these artificial opals have been used to prepare inverse opal photonic crystals in order to get higher refractive index contrast by infiltration of the inorganic compounds and removal of the polymer spheres. Although the self-organization based approach is simple and cost-effective, the lattice symmetry is limited to close-packed lattice inherent in opals, which is also not the optimal lattice structure for the first order complete photonic bandgap opening in three-dimensional photonic crystals [21]. Although several methods have been demonstrated to prepare three-dimensional photonic crystals [22–26], fabrication of high-quality three-dimensional photonic crystals still remains a challenge.

In the last fifteen years, the most studied fabrication process for photonic crystal devices is based on two-dimensional pattering of semiconductor (Si or III-V compounds) slabs using electron beam lithography in combination with dry etching technique. T. Krauss et al. (Univ. St. Andrews) demonstrated a first two-dimensional photonic bandgap structure of photonic crystals at optical wavelength in 1996 [27]. State of the art nanofabrication technology by borrowing the methods used in the semiconductor industry allows fabrication of two-dimensional photonic crystals with sufficient precision to prevent scattering losses. Electron beam lithography technique uses electron beam with energies typically from 30 to 125 keV to form patterns on a surface covered by a thin layer of a spin-on e-beam resist. The primary advantage of this technique over conventional photolithography is that electrons are orders of magnitude smaller than photons, so it is one of the ways to beat the diffraction limit of light and make features in the nanometer regime. Typical e-beam resist patterns of two-dimensional photonic crystal structure are shown in Fig. 2.2. The electron beam is scanning across the surface to be patterned point by point in the writhing field (known as vector scan). This makes the method time-consuming, which is the key limitation of electron beam lithography. Subsequent stage in the fabrication process is dry etching for high-resolution pattern transfer. In this process, a plasma chemical removes certain areas of a material surface that are not protected by e-beam

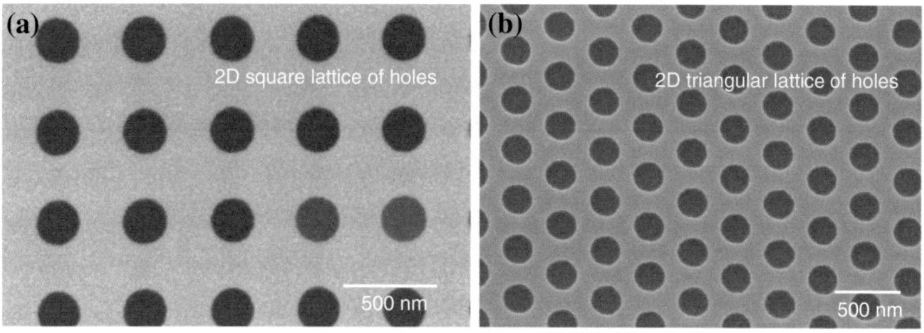

Fig. 2.2 Scanning electron micrograph of typical resist patterns of (**a**) 2D square lattice and (**b**) triangular lattice photonic crystals fabricated by e-beam lithography

resist. High selectivity and high-aspect ratio etching requires high-density plasma processing. Inductively Coupled Plasma (ICP) method provides a dense plasma with an electron density up to $10^{12}\,cm^{-3}$ which is a 5 to 10 times higher plasma density than in traditional capacitively coupled Reactive Ion Etching (RIE) generator. The ICP reactor is the plasma chamber using a 13.56 MHz radio-frequency power source connected to a single-turn or helical coil antenna through a matching box and coupled via a quartz or ceramic window at top or side of the chamber. The etching sample is placed on a cooled stage in the chamber and biased by rf (13.56 MHz) power to provide ion energy. The advantage of this configuration is the ability to form very dense plasmas and the possibility of independent control of the energy of ion bombardment and the plasma density. At the same time, the chamber can be kept lower pressure (below a few mTorr) in compared with several hundred mTorr in the conventional RIE reactor, to avoid increased scattering, whereby the directionality of the ion bombardment is maintained. Moreover, ICP etching method uses gas chemistries in order to obtain highly anisotropic etch profiles. Patterned materials are etched by combination of an etching chemistry and a passivating chemistry. The passivation chemistries supply the deposited polymer inhibitors on sidewalls to protect vertical sidewalls of a patterned material from chemical isotropic etching. For photonic applications, it is seriously important to reduce the sidewall roughness that contributes to low scattering losses of fabricated devices. The two-dimensional photonic crystal waveguides fabricated using these techniques exhibited high precision features on the level of single nm RMS in recent years. Figure 2.3 shows examples of our etching results of the two-dimensional photonic crystal slab waveguides fabricated using the optimal ICP parameters for producing straight sidewalls.

Recently, nanoimprint technology has been developed, which provides a cheaper route towards commercial fabrication of planar nanometer scale patterns. Thermal nanoimprint lithography deforms soft materials by heating them above the glass transition temperature Tg, and hardens the materials by cooling [28–32]. The nanoimprint lithography offers extreme nano-structuring capability with the promise of low cost, high throughput and high volume fabrication. The spatial resolution in the structure patterned by this technique is around 5 to 15 nm [31, 33]. Recently, nanoimprint

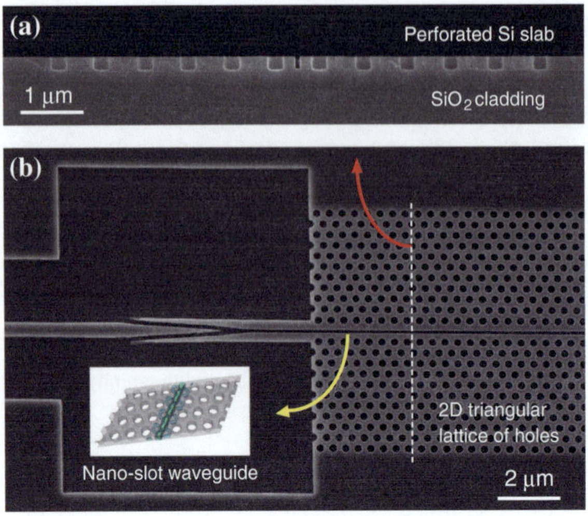

Fig. 2.3 SEM micrographs of (**a**) cross-sectional and (**b**) top views of the Si slotted photonic crystal waveguide fabricated by the ICP dry-etching method. The inset shows a calculated optical mode profile

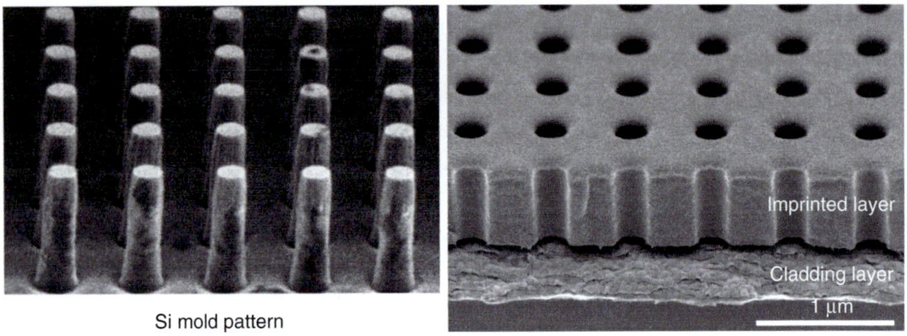

Fig. 2.4 Cross-sectional SEM images of a Si mold with square lattice pillar pattern (*left*) and a patterned polymer photonic crystal structure fabricated by nanoimprint lithography technique (*right*)

machines for patterning over a diameter of 300 mm have became commercially available. Example of the imprinted periodic lattice structure comprised of air holes are shown in Fig. 2.4. It is expected that nanoimprint technology will be a very useful method to pattern photonic crystal structures for practical purposes and industrial use.

2.3 Theoretical and Experimental Photonic Band Structure

Gaining an understanding of the theoretical and experimental photonic band structure and how it affects variety of optical phenomena (both linear and nonlinear) is critical to advancing both fundamental research and technological applications of photonic crystals, since many photonic properties of photonic crystals and their underlying physical mechanisms are closely related to the shape of the photonic band structure.

2.3.1 Fundamental Theory and Plane-Wave Expansion Method

The photonic band structure is a band dispersion relation between frequency ω and wavevector k. The plane-wave expansion method is a basic technique for the photonic band structure calculations, which has made an important contribution to the development of initial photonic crystal works [9, 34–38]. In addition, the correspondence relations between the Schrödinger equation for electrons and the wave equation for the photons in this method have played an important role for understanding of the novel optical functions of a photonic crystal. This numerical method is based on the Fourier expansion of the electromagnetic field and the dielectric function. We start with Maxwell's equations, because all of macroscopic electromagnetism, including the propagation of light in a photonic crystal, is governed by these equations. In MKS units, they are

$$\nabla \cdot \{\epsilon(\boldsymbol{r})\boldsymbol{E}(\boldsymbol{r},t)\} = 0 \tag{2.1}$$

$$\nabla \cdot \boldsymbol{H}(\boldsymbol{r},t) = 0 \tag{2.2}$$

$$\nabla \times \boldsymbol{E}(\boldsymbol{r},t) = -\mu_0 \frac{\partial}{\partial t}\boldsymbol{H}(\boldsymbol{r},t) \tag{2.3}$$

$$\nabla \times \boldsymbol{H}(\boldsymbol{r},t) = \epsilon_0\epsilon(\boldsymbol{r}) \frac{\partial}{\partial t}\boldsymbol{E}(\boldsymbol{r},t) \tag{2.4}$$

where $\boldsymbol{E}$ and $\boldsymbol{H}$ are electric and magnetic fields, and we assume no free charges and currents. When we eliminate $\boldsymbol{E}$ or $\boldsymbol{H}$ from these equations, we obtain the equations satisfied by $\boldsymbol{E}$ or $\boldsymbol{H}$, which we write in the forms

$$\frac{1}{\epsilon(\boldsymbol{r})}\nabla \times \{\nabla \times \boldsymbol{E}(\boldsymbol{r},t)\} = -\frac{1}{c^2}\frac{\partial^2}{\partial t^2}\boldsymbol{E}(\boldsymbol{r},t) \tag{2.5}$$

$$\nabla \times \left\{\frac{1}{\epsilon(\boldsymbol{r})}\nabla \times \boldsymbol{H}(\boldsymbol{r},t)\right\} = -\frac{1}{c^2}\frac{\partial^2}{\partial t^2}\boldsymbol{H}(\boldsymbol{r},t) \tag{2.6}$$

These are the wave equations, where c stands for the light velocity in the vacuum,

$$c = \frac{1}{\sqrt{\epsilon_0\mu_0}} \tag{2.7}$$

In general, the electric and magnetic fields are functions of time and space. However, because the Maxwell equations are linear, we can separate out the time dependence by expanding the fields into a set of harmonic modes,

$$\boldsymbol{E}(\boldsymbol{r},t) = e^{-i\omega t}\boldsymbol{E}(\boldsymbol{r}) \tag{2.8}$$

$$\boldsymbol{H}(\boldsymbol{r},t) = e^{-i\omega t}\boldsymbol{H}(\boldsymbol{r}) \tag{2.9}$$

To seek the solutions for the mode profiles of a given eigen-angular frequency, we insert the above equations into (2.5) and (2.6):

$$\Theta_E \boldsymbol{E} \equiv \frac{1}{\epsilon(\boldsymbol{r})}\nabla \times \{\nabla \times \boldsymbol{E}(\boldsymbol{r})\} = \frac{\omega^2}{c^2}\boldsymbol{E}(\boldsymbol{r}) \tag{2.10}$$

$$\Theta_H \boldsymbol{H} \equiv \nabla \times \left\{\frac{1}{\epsilon(\boldsymbol{r})}\nabla \times \boldsymbol{H}(\boldsymbol{r})\right\} = \frac{\omega^2}{c^2}\boldsymbol{H}(\boldsymbol{r}) \tag{2.11}$$

where Θ_E and Θ_H are identified as the differential operators. This situation arises often in mathematical physics, and is called a "Hermitian" eigenvalue problem. If the result of an operation on a function is just the function itself, multiplied by some constant, then the function is called an eigenfunction or eigenvector of that operator, and the multiplicative constant is called the eigenvalue. Here, the eigenfunctions $\boldsymbol{E}(\boldsymbol{r})$ or $\boldsymbol{H}(\boldsymbol{r})$ are the electric or magnetic field patterns of the harmonic modes, and the eigenvalues $(\omega/c)^2$ are proportional to the squared frequencies of those modes. In photonic crystals, $\epsilon(\boldsymbol{r})$ is a periodic function and satisfies the relation,

$$\epsilon(\boldsymbol{r}+\boldsymbol{R}) = \epsilon(\boldsymbol{r}) \tag{2.12}$$

where the set of vectors $\boldsymbol{R}$ is called the lattice vectors of the photonic crystal lattice. The lattice vectors $\boldsymbol{R}$ can be written in terms of the primitive lattice vectors $\boldsymbol{a}_1$, $\boldsymbol{a}_2$, and $\boldsymbol{a}_3$. Therefore, we expand the periodic dielectric function in a Fourier series,

$$\frac{1}{\epsilon(\boldsymbol{r})} = \sum_{\boldsymbol{G}} \kappa(\boldsymbol{G}) e^{i\boldsymbol{G}\boldsymbol{r}} \tag{2.13}$$

where the vectors $\boldsymbol{G}$ are the reciprocal lattice vectors,

$$\boldsymbol{G} = l\boldsymbol{b}_1 + m\boldsymbol{b}_2 + n\boldsymbol{b}_3 \tag{2.14}$$

Here $\boldsymbol{b}_1$, $\boldsymbol{b}_2$, and $\boldsymbol{b}_3$ are the primitive reciprocal lattice vectors and l, m, and n are integers. The primitive vectors obey the following orthogonality relation:

$$\boldsymbol{a}_i \cdot \boldsymbol{b}_j = 2\pi\delta_{ij} \tag{2.15}$$

where δ_{ij} is Kronecker's delta. Because of the periodicity of $\epsilon(\boldsymbol{r})$, we can apply Bloch's theorem to $E(\boldsymbol{r})$ and $H(\boldsymbol{r})$. The electric and magnetic fields are thus characterized by a wave vector $\boldsymbol{k}$ in the first Brillouin zone and expressed as

$$E(\boldsymbol{r}) = e^{i\boldsymbol{k}\cdot\boldsymbol{r}}\boldsymbol{u}_{\boldsymbol{k}}(\boldsymbol{r}) \tag{2.16}$$

$$H(\boldsymbol{r}) = e^{i\boldsymbol{k}\cdot\boldsymbol{r}}\boldsymbol{v}_{\boldsymbol{k}}(\boldsymbol{r}) \tag{2.17}$$

$$\boldsymbol{u}_{\boldsymbol{k}}(\boldsymbol{r}+\boldsymbol{a}_i) = \boldsymbol{u}_{\boldsymbol{k}}(\boldsymbol{r}) \tag{2.18}$$

$$\boldsymbol{v}_{\boldsymbol{k}}(\boldsymbol{r}+\boldsymbol{a}_i) = \boldsymbol{v}_{\boldsymbol{k}}(\boldsymbol{r}) \tag{2.19}$$

where $\boldsymbol{u}_{\boldsymbol{k}}(\boldsymbol{r})$ and $\boldsymbol{v}_{\boldsymbol{k}}(\boldsymbol{r})$ are periodic vectorial functions. Because of the spatial periodicity of these functions, they can be expanded in terms of an infinite discrete sum of spatial harmonics. Accordingly, the electric and magnetic fields in a periodic system can be written as:

$$\boldsymbol{E}_{\boldsymbol{k}}(\boldsymbol{r}) = \sum_{\boldsymbol{G}} \boldsymbol{E}_{\boldsymbol{k}}(\boldsymbol{G})e^{i(\boldsymbol{k}+\boldsymbol{G})\cdot\boldsymbol{r}} \tag{2.20}$$

$$\boldsymbol{H}_{\boldsymbol{k}}(\boldsymbol{r}) = \sum_{\boldsymbol{G}} \boldsymbol{H}_{\boldsymbol{k}}(\boldsymbol{G})e^{i(\boldsymbol{k}+\boldsymbol{G})\cdot\boldsymbol{r}} \tag{2.21}$$

Substituting (2.13), (2.20), and (2.21) into the wave equations (2.10) and (2.11), we arrive at the following eigenvalue equations in terms of the Fourier components of the electric or magnetic fields:

$$-\sum_{\boldsymbol{G}'} \kappa(\boldsymbol{G}-\boldsymbol{G}')(\boldsymbol{k}+\boldsymbol{G}') \times [(\boldsymbol{k}+\boldsymbol{G}') \times \boldsymbol{E}_{\boldsymbol{k}}(\boldsymbol{G}')] = \frac{\omega_k^2}{c^2}\boldsymbol{E}_{\boldsymbol{k}}(\boldsymbol{G}) \tag{2.22}$$

$$-\sum_{\boldsymbol{G}'} \kappa(\boldsymbol{G}-\boldsymbol{G}')(\boldsymbol{k}+\boldsymbol{G}) \times [(\boldsymbol{k}+\boldsymbol{G}') \times \boldsymbol{H}_{\boldsymbol{k}}(\boldsymbol{G}')] = \frac{\omega_k^2}{c^2}\boldsymbol{H}_{\boldsymbol{k}}(\boldsymbol{G}) \tag{2.23}$$

These equations present a standard eigenvalue problem with respect to the value of the eigen-angular frequency $\omega_{\boldsymbol{k}}$ for a given value of a wave vector $\boldsymbol{k}$. By solving numerically above eigenvalue problem for each $\boldsymbol{k}$ in the first Brillouin zone, we can obtain the photonic band structure in photonic crystals. In actual calculations, instead of an infinite number of Fourier coefficients, a finite number N of $\boldsymbol{G}'$ (i.e., the number of plan waves) is used. To achieve adequate accuracy in plane-wave expansion method, a sufficiently large number N is required. In particular, in the case of a large scale structure such as a supercell of a photonic crystal with defects, several thousands of N are needed, which is time consuming. For the comprehensive theory of the plane-wave expansion method including the convergence problem of the numerical calculations, the reader is referred to the appropriate references [39–41].

2.3.2 Finite Difference Time Domain Method

Finite difference time domain (FDTD) is a powerful numerical method that calculates directly the time-dependent Maxwell's curl equations in the analytical divided finite real space. Although FDTD method requires large amounts of memory and long computing time, this method has been widely applied to model interaction of electromagnetic waves in a broad range of photonic crystal structures because of the remarkable progress in recent computer technology. Since the FDTD method is a rigorous and direct solution of Maxwell's curl equations, it does not have any approximations or theoretical restrictions and includes many more effects than other approximate methods, such as the plane-wave expansion method. In this subsection, we describe the algorism and the computational techniques of the three-dimensional FDTD method for the electromagnetic fields and the photonic band structure calculations, considering the imaginary part of the dielectric function of the materials.

The time-development Maxwell's equations can be written in the following form

$$\nabla \times \mathbf{E}(\mathbf{r}, t) = -\mu(\mathbf{r})\frac{\partial}{\partial t}\mathbf{H}(\mathbf{r}, t) \tag{2.24}$$

$$\nabla \times \mathbf{H}(\mathbf{r}, t) = \sigma(\mathbf{r})\mathbf{E} + \epsilon(\mathbf{r})\frac{\partial}{\partial t}\mathbf{E}(\mathbf{r}, t) \tag{2.25}$$

where $\epsilon(\boldsymbol{r})$, $\mu(\boldsymbol{r})$, and $\sigma(\boldsymbol{r})$ are the permittivity, permeability, and conductivity, respectively, and are functions of the spatial coordinates. These equations can be discretized in space and time by a Yee-cell algorithm [42, 43]. The Yee-algorithm solves for both electric and magnetic fields using the coupled Maxwell's curl equations. By using both $\boldsymbol{E}$ and $\boldsymbol{H}$ information instead of using only the $\boldsymbol{E}$ or $\boldsymbol{H}$ field as happens for the wave equation, this scheme is very robust, versatile and widely applicable for a broad class of material structures. Arrangement of the six electromagnetic field components on the Yee cell in a three-dimensional space are shown in Fig. 2.5. Each single component of a field is surrounded by four components of the other field. The electric and magnetic field components are also offset in the discretized time domain by $1/2$ time step in an alternate manner called a leapfrog arrangement. The leapfrog algorithm is fully explicit, which avoids problems involved with simultaneous equations and matrix inversion. In this algorithm, the electric and magnetic field components are updated in discrete time steps. This process permits the numerical wave to propagate in the computational mesh as the simulation continues until time-stepping is concluded. In the Yee-cell with the leapfrog time-stepping algorithm, the finite-difference equations for both the space and time derivatives are central-difference, which permits second-order accuracy to be achieved in the space and time increments.

The finite-difference expressions for Maxwell's curl equations in three-dimension is obtained by applying the above ideas and process to Eqs. (2.24) and (2.25). We now write out the vector components of the curl operators of (2.24) and (2.25) in Cartesian coordinates. This also can be expressed by six coupled scalar equations.

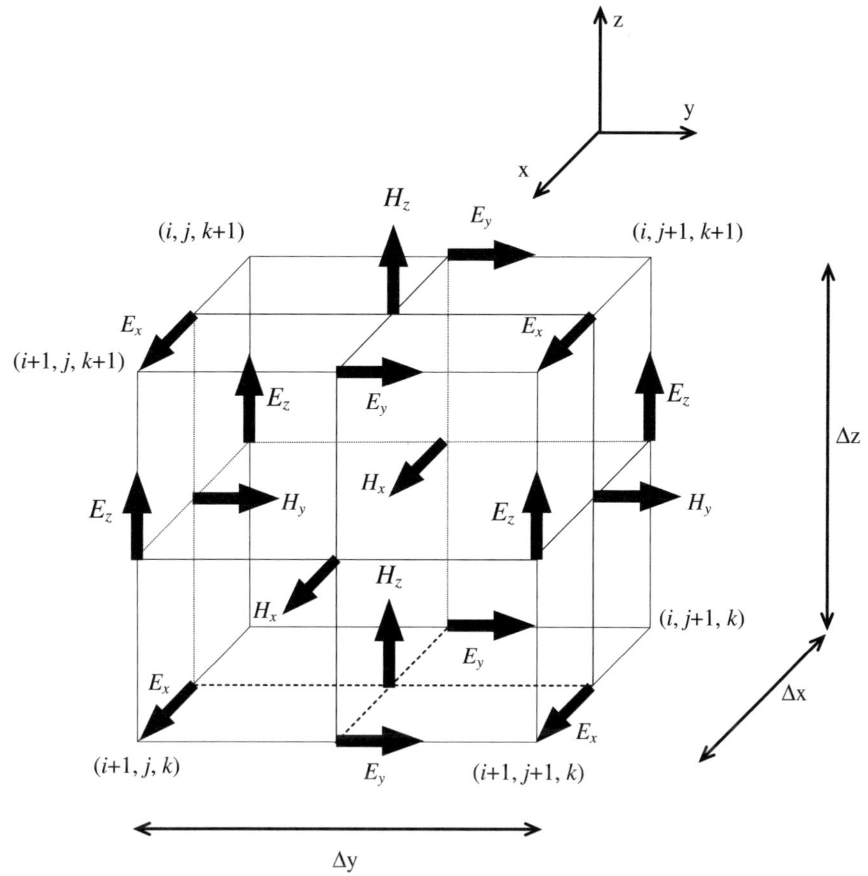

Fig. 2.5 The location of the electric and magnetic field vector components in the three-dimensional Yee cell

For example we start by considering the Ex field component,

$$\frac{\partial E_x}{\partial t} = \frac{1}{\epsilon}\left(\frac{\partial H_z}{\partial y} - \frac{\partial H_y}{\partial z} - \sigma E_x\right) \tag{2.26}$$

This equation is then discretized using central differences to the time and space derivatives following the Yee-cell and leapfrog scheme by referring to Fig. 2.5. Consequently, we obtain the desired explicit time-stepping relation for E_x,

$$E_x^n(i+\frac{1}{2},j,k) = \frac{1-\frac{\sigma(i+\frac{1}{2},j,k)\Delta t}{2\epsilon(i+\frac{1}{2},j,k)}}{1+\frac{\sigma(i+\frac{1}{2},j,k)\Delta t}{2\epsilon(i+\frac{1}{2},j,k)}} E_x^{n-1}\left(i+\frac{1}{2},j,k\right)$$
$$+\frac{\frac{\Delta t}{\epsilon}(i+\frac{1}{2},j,k)}{1+\frac{\sigma(i+\frac{1}{2},j,k)\Delta t}{2\epsilon(i+\frac{1}{2},j,k)}} \frac{H_z^{n-\frac{1}{2}}(i+\frac{1}{2},j+\frac{1}{2},k)-H_z^{n-\frac{1}{2}}(i+\frac{1}{2},j-\frac{1}{2},k)}{\Delta y}$$
$$-\frac{\frac{\Delta t}{\epsilon}(i+\frac{1}{2},j,k)}{1+\frac{\sigma(i+\frac{1}{2},j,k)\Delta t}{2\epsilon(i+\frac{1}{2},j,k)}} \frac{H_y^{n-\frac{1}{2}}(i+\frac{1}{2},j+\frac{1}{2},k)-H_y^{n-\frac{1}{2}}(i+\frac{1}{2},j-\frac{1}{2},k)}{\Delta z} \tag{2.27}$$

where Δt is the time step, and Δy and Δz are the space increments along the y- and z-axes, respectively. For the E_y and E_z field components, the time-stepping discretized finite-difference expressions can be obtained by applying Yee algorithm to Maxwell's equations in the same manner,

$$E_y^n(i,j+\frac{1}{2},k) = \frac{1-\frac{\sigma(i,j+\frac{1}{2},k)\Delta t}{2\epsilon(i,j+\frac{1}{2},k)}}{1+\frac{\sigma(i,j+\frac{1}{2},k)\Delta t}{2\epsilon(i,j+\frac{1}{2},k)}} E_y^{n-1}\left(i,j+\frac{1}{2},k\right)$$
$$+\frac{\frac{\Delta t}{\epsilon}(i,j+\frac{1}{2},k)}{1+\frac{\sigma(i,j+\frac{1}{2},k)\Delta t}{2\epsilon(i,j+\frac{1}{2},k)}} \frac{H_x^{n-\frac{1}{2}}(i,j+\frac{1}{2},k+\frac{1}{2})-H_x^{n-\frac{1}{2}}(i,j+\frac{1}{2},k-\frac{1}{2})}{\Delta z}$$
$$-\frac{\frac{\Delta t}{\epsilon}(i,j+\frac{1}{2},k)}{1+\frac{\sigma(i,j+\frac{1}{2},k)\Delta t}{2\epsilon(i,j+\frac{1}{2},k)}} \frac{H_z^{n-\frac{1}{2}}(i+\frac{1}{2},j+\frac{1}{2},k)-H_z^{n-\frac{1}{2}}(i-\frac{1}{2},j-\frac{1}{2},k)}{\Delta x} \tag{2.28}$$

and

$$E_z^n(i,j,k+\frac{1}{2}) = \frac{1-\frac{\sigma(i,j,k+\frac{1}{2})\Delta t}{2\epsilon(i,j,k+\frac{1}{2})}}{1+\frac{\sigma(i,j,k+\frac{1}{2})\Delta t}{2\epsilon(i,j,k+\frac{1}{2})}} E_z^{n-1}\left(i,j,k+\frac{1}{2}\right)$$
$$+\frac{\frac{\Delta t}{\epsilon}(i,j,k+\frac{1}{2})}{1+\frac{\sigma(i,j,k+\frac{1}{2})\Delta t}{2\epsilon(i,j,k+\frac{1}{2})}} \frac{H_y^{n-\frac{1}{2}}(i+\frac{1}{2},j,k+\frac{1}{2})-H_y^{n-\frac{1}{2}}(i-\frac{1}{2},j,k+\frac{1}{2})}{\Delta x}$$
$$-\frac{\frac{\Delta t}{\epsilon}(i,j+\frac{1}{2},k)}{1+\frac{\sigma(i,j+\frac{1}{2},k)\Delta t}{2\epsilon(i,j+\frac{1}{2},k)}} \frac{H_y^{n-\frac{1}{2}}(i,j+\frac{1}{2},k+\frac{1}{2})-H_y^{n-\frac{1}{2}}(i,j-\frac{1}{2},k+\frac{1}{2})}{\Delta y} \tag{2.29}$$

Similarly, we can also derive the time-stepping discretized finite-difference expressions for H_x, H_y and H_z. Three equations for H_x, H_y and H_z have a form similar to that of the E equations above. The detailed derivation of these equations for the field components will not be repeated here, but the reader is referred to the appropriate references [43–48].

In above updating process, needless to say, propagating of wave modes in the numerical mesh does not occur in the absence of an initial field. Therefore, in the photonic band structure calculations of photonic crystals, the initial TE- or TM-polarized electric field that has the distribution with low symmetry in the photonic crystal model structure is used to excite the TE- or TM-like modes. The initial electromagnetic field evolves in time by the discrete Yee-cell and leapfrog scheme until a steady state is reached or the fixed maximum number of time steps is reached. Here, the choice of the space and time increments can affect the propagation characteristics of numerical waves in the Yee computational domain, and consequently the numerical accuracy of this scheme. Typically, the space increment must be able to resolve the wavelength in time, and therefore usually be less than $\lambda/10$ to $\lambda/40$ where λ is the wavelength in the materials. However, since the numerical error depends on a shape of model structures, the convergence behavior for the size of Yee-cell dividing space should be examined. Moreover, since the discrete Yee-cell and leapfrog scheme is based in the time domain, spatial step parameters relate to the temporal grid. In order to obtain a numerical stability, Δt must also be bounded to ensure the following Courant condition,

$$v_{\max}\Delta t \leq \left(\frac{1}{\Delta x^2}+\frac{1}{\Delta y^2}+\frac{1}{\Delta z^2}\right)^{-\frac{1}{2}} \tag{2.30}$$

where $v_{\max}$ is the maximum phase velocity of the electromagnetic wave in this analytical model.

To simulate the photonic band structure of the two-dimensional photonic crystals in the three-dimensional computational mesh, the Bloch boundary condition is applied to the side boundaries normal to plane of the slab waveguides, and is given as follows,

$$\boldsymbol{E}(\boldsymbol{r}+\boldsymbol{a}_i,t)=\boldsymbol{E}(\boldsymbol{r},t)e^{i\boldsymbol{k}\boldsymbol{a}_i} \tag{2.31}$$

$$\boldsymbol{H}(\boldsymbol{r}+\boldsymbol{a}_i,t)=\boldsymbol{H}(\boldsymbol{r},t)e^{i\boldsymbol{k}\boldsymbol{a}_i} \tag{2.32}$$

The Bloch boundary conditions over a unit cell of a photonic crystal give the k vector dependence of the eigenmode. Mur's absorbing boundary condition [49, 50] or Berenger's perfect matched layer (PML) condition [51, 52] for absorbing boundary conditions so as to negate a reflection is applied to the top and bottom boundaries. All the fields are obtained in the time domain in this method. In order to obtain the spectral information, the calculated fields need to be converted from the time domain into the frequency domain by taking their Fourier transform. The peaks of the spectral intensity correspond to the locations of the eigen-photon-energies of the photonic band modes.

2.3.3 *Direct Determination of Experimental Photonic Band Structure*

In order to design high-efficiency photonic applications and to achieve a proper understanding of the unique optical phenomena in photonic crystals, it is essential to engineer the optical dispersion relation and to study the experimental photonic band structure and the key fundamental characteristics such as the group velocity dispersion characteristics.

The experimental photonic band structure of the photonic crystal waveguide can be determined from the incident angle dependence of the wavelength positions of sharp resonance dips in the polarized-angular-dependent reflectance spectra [53–55]. The sharp dips in the reflectivity spectra originate from surface coupling between external free photons and in-plane photonic band modes in the waveguide at resonance energies and in-plane wave vectors [54]. Therefore these dips provide information about the photonic band dispersion, but not about the photonic band gap and its angular dependence on the Bragg peak [56].

However, the photonic bands that can be probed with this technique are limited to those above the light cone. The light line in a vacuum ($\omega = ck$) separates the observable region ($\omega > ck$), in which the modes are oscillatory in air, from the region in which modes are evanescent in air and cannot couple with external free photons ($\omega < ck$). However, engineering applications of two-dimensional photonic crystal waveguides require the use of modes below the light line due to their vertical confinement and long lifetimes. The dispersion characteristics of waveguiding modes below the light line have been studied indirectly by interference measurements of transmission in a 2D photonic crystal slab [57]. This method can deduce the group index (n_g) dispersion, but it cannot accurately determine the shape of the band dispersion due to the uncertainty in the origin position of the k-axis. Although determining the photonic band dispersion relation of the waveguiding modes below the light line is indispensable for understanding the unique properties of photonic crystal waveguides, experimental information on the specific shape of these band dispersions has scarcely been obtained directly.

In this subsection, we demonstrate the direct determination of the experimental photonic band structure below the light line in a two-dimensional photonic crystal waveguide using angle-resolved attenuated total reflection spectroscopy of a prism coupling arrangement over a wide frequency range. By this technique, we can observe sharp dips in the reflectance spectra originating from resonance coupling between the external evanescent wave from total reflection off the prism and the waveguiding modes in the photonic crystal waveguide. This provides clear information on individual band components by resolving the angle (i.e., wave vector k) and photon energy.

Because waveguiding modes in 2D photonic crystals with $k > \omega/c$ cannot be matched to propagating modes incident from free space, they cannot be detected by external plane-wave excitation. To examine the experimental photonic band structure of these modes, the incident radiation is coupled to waveguiding modes using a coupling prism in the Otto configuration [58]. The experimental geometry is shown in Fig. 2.6. A beam of radiation is incident from air through one face of a high-refractive-index prism. The beam passes through the prism and is incident on a second interface at an angle. For angles greater than a critical angle, the beam undergoes total internal reflection at the second interface and exits the prism at the third interface. At the surface of the reflection interface, a decaying evanescent wave exists on the air side of the prism, for which the parallel component of the wave vector satisfies $k_{\parallel} > \omega/c$. When the surface of the 2D photonic crystal waveguide is brought close to (but not touching) the prism, this evanescent wave can couple to the guiding mode. A sharp dip originating from resonance between the evanescent wave and the waveguiding mode can be detected in the reflectance spectrum.

As an example, Fig. 2.7 shows typical angle-dependent attenuated total reflection spectra of the 2D photonic crystal slab waveguide along the $\Gamma - X$ lattice direction, taken with TE polarization. Several sharp dips are clearly observable, demonstrating that the evanescent field above the prism resonantly couples to in-plane photonic band modes in the 2D photonic crystal. The observed resonance wavelengths clearly depend on the incident angle, which determines the in-plane wave vectors, $k_{\parallel} = (2\pi/\lambda)n_p \sin\theta_{\text{int}}$, where λ is the wavelength of the incident light and $\theta_{\text{int}} = 0°$ is normal incidence. The resonance dip at each resonance angle in the reflectivity is

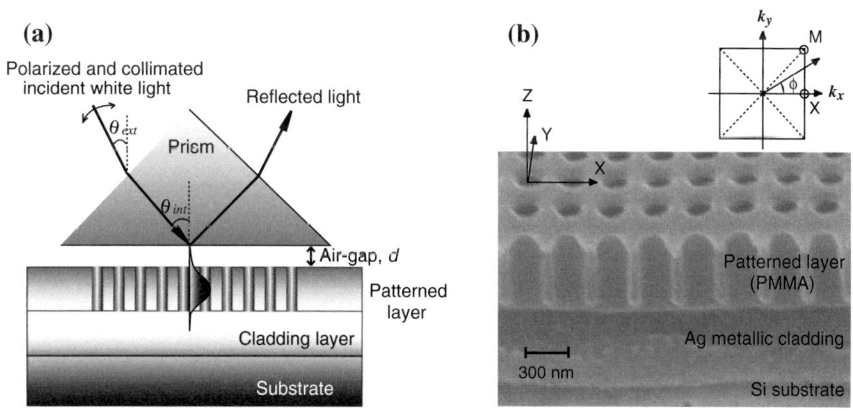

Fig. 2.6 **a** Schematic diagram showing the experimental geometry of angle-resolved attenuated total reflection spectroscopy. The coupling prism in the Otto configuration is used to couple incident radiation to waveguiding modes in the 2D PhC. **b** Example of the 2D PhC waveguide used in the experiments. The inset shows the corresponding 2D first Brillouin zone and the high-symmetry lattice points

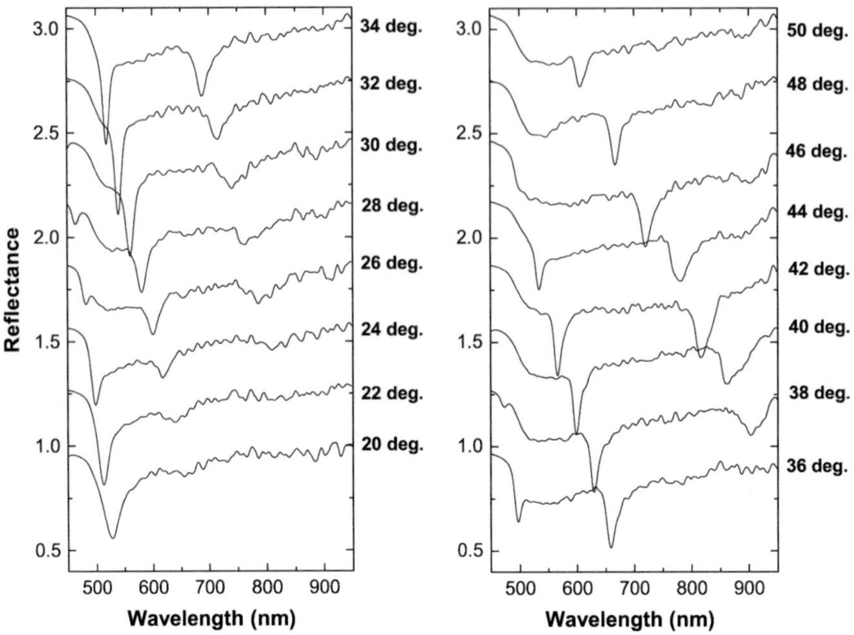

Fig. 2.7 Typical measured reflectance spectra for various angles of incidence with TE polarization along the $\Gamma - X$ direction. For clarity, the spectra are shifted along the vertical axis by 0.3 with respect to each other

thus directly connected to the experimental dispersion curves, i.e., the experimental photonic band structure.

Figure 2.8 shows an example of experimental photonic band structure of the photonic crystal along the $\Gamma - X$ line obtained from the angle-dependent attenuated total reflection spectra, compared with the theoretical band structure calculated by the 3D-FDTD method for a model structure identical to the experimental sample structure. The straight solid line corresponds to the light line in a vacuum ($\omega = ck$) and separates the region in which the modes are oscillatory in air ($\omega > ck$) from the region in which modes are evanescent in air ($\omega < ck$). In a standard angular-dependent reflectivity measurement without a coupling prism, the experimental bands that can be probed are limited to those above the light cone. In angle-resolved attenuated total reflection spectroscopy, the in-plane wave vectors $k_{\|}$ of the incident plane-wave are increased in proportion to the prism index n_p, and hence many of the photonic band dispersions under the straight dashed line correspond to the boundary of the enlarged light cone in the prism ($\omega > ck/n_p$). The features of the experimental band structure are in reasonably good agreement with the theoretical curves. This agreement demonstrates that it is realistically possible to design advanced fundamental research and technological applications of photonic crystal slabs using this technology, since many optical properties of the waveguiding modes are closely related to the shape of the band structure under the light line in air.

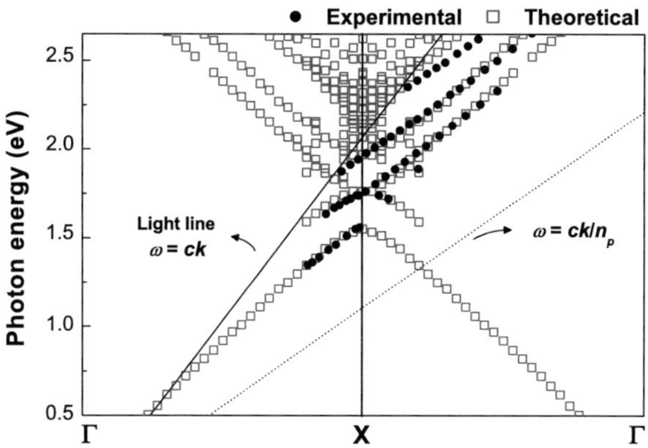

Fig. 2.8 Experimental photonic band structure under the light line of the 2D photonic crystal waveguide along the Γ-X line obtained by polarized angular-dependent reflection spectroscopy (*closed circles*) and the theoretical band structure calculated by the 3D FDTD method (*open squares*). No empirical adjustments were made in the calculations

It has been also proved that the fluctuation in the photon energy domain between the band modes of the photonic crystal with the prism and the isolated photonic crystal are barely significant ($\Delta h\nu / h\nu < 0.6\,\%$) even at the well-coupled separation distance d of 300 nm between the photonic crystal and prism [59]. These results clearly demonstrate that, except in the case of over coupling ($d < 100$ nm), the angle-resolved attenuated total reflection spectroscopy technique is accurate and suitable for directly determining the experimental photonic band structure below the light line in 2D photonic crystal waveguides.

2.4 Enhancement of Nonlinear Optical Processes

Photonic crystals are very attractive materials for use in the propagation of electromagnetic radiation and for their ability to control the band dispersion characteristics, as described in Sect. 2.3. These characteristics have opened up unprecedented and exciting possibilities in a wide range of photonic information technologies, including photonic circuit devices [60–62], controlling spontaneous emission [63–65], microcavity works [66–68], negative refraction [69–72], and diverse applications [73–77]. For example, in line-defect photonic crystal waveguides, a waveguiding mode located in a photonic band gap can produce a wavelength-scale sharp bend in the light propagation [60, 62]. Also, specific features of the band dispersion relation and conservation of the tangential component of the wave vector at the boundary in photonic crystals enable the realization of super-prism, focusing, and negative refraction effects [69–72, 78, 79]. Furthermore, photonic band structure features and large group velocity dispersion characteristics in photonic crystals are expected to find use in new active and novel high-efficiency nonlinear optical applications such as high-efficiency optical switching [80–83], frequency conversion devices [84–88], and others [89–91], in combination with nonlinear optical host materials. These applications are possible with photonic crystals because an extremely slow group velocity, originating from anomalous band dispersions at the band edge and/or a very flat band, produces a strong enhancement in the electromagnetic field of the excitation wave. We finish this chapter by briefly showing the exciting possibilities of a photonic crystal in nonlinear optics.

2.4.1 Engineered Third-Order Nonlinear Optical Responses

Interest has grown rapidly in photonic crystals exhibiting nonlinear optical switching effects as such materials are strong candidates for practical all-optical signal processing devices [80–83]. The optical nonlinearities of conventional nonlinear materials are usually very weak. However, tailoring of the band dispersions in photonic crystal systems can produce large nonlinearities and minimize the power requirements for switching processes. This arises from the anomalous dispersion characteristics of photonic crystals, which enhance the electromagnetic field due to their extremely slow group velocities as compared with conventional uniform materials. In this regard, a two-dimensional (2D) photonic crystal waveguide represents one of the most promising photonic crystal structures for efficient nonlinear applications. Unique band dispersions can be designed through control of the 2D photonic crystal configuration and a high optical intensity can be maintained over a long interaction distance.

Clarifying the relation between the nonlinear optical response and the nature of the band dispersion at the resonance state, and engineering the matter-radiation nonlinear interaction by controlling the band dispersion are essential for the realization of desirable nonlinear applications, such as practical all-optical switching and processing devices with very low operational power and ultra-small dimensions, which are greatly demanded by photonic information technology. In this subsection, we experimentally identify nonlinear optical changes arising from modification of the photonic band structure in a 2D photonic crystal waveguide with a Kerr nonlinearity.

To examine the nonlinear optical changes arising from modifications of the photonic bands driven by an external laser field resonantly coupled to the photonic bands, linear and nonlinear angle-resolved reflectivity measurements are performed. The examples of the linear and nonlinear angle-dependent reflectivities of the 2D photonic crystal waveguide composed of the nonlinear optical polymer DR1/PMMA are shown in Fig. 2.9. These measurements provide a sharp distinction between the different photonic bands by separating the linear components in the angle (or wave vector, k) domain and also convey information on the nonlinear optical responses by monitoring the distinct dynamics of each photonic band in the 2D photonic crystal system. Furthermore, because this technique can separate photonic bands of the same energy, we can investigate the nonlinear optical changes of the different photonic

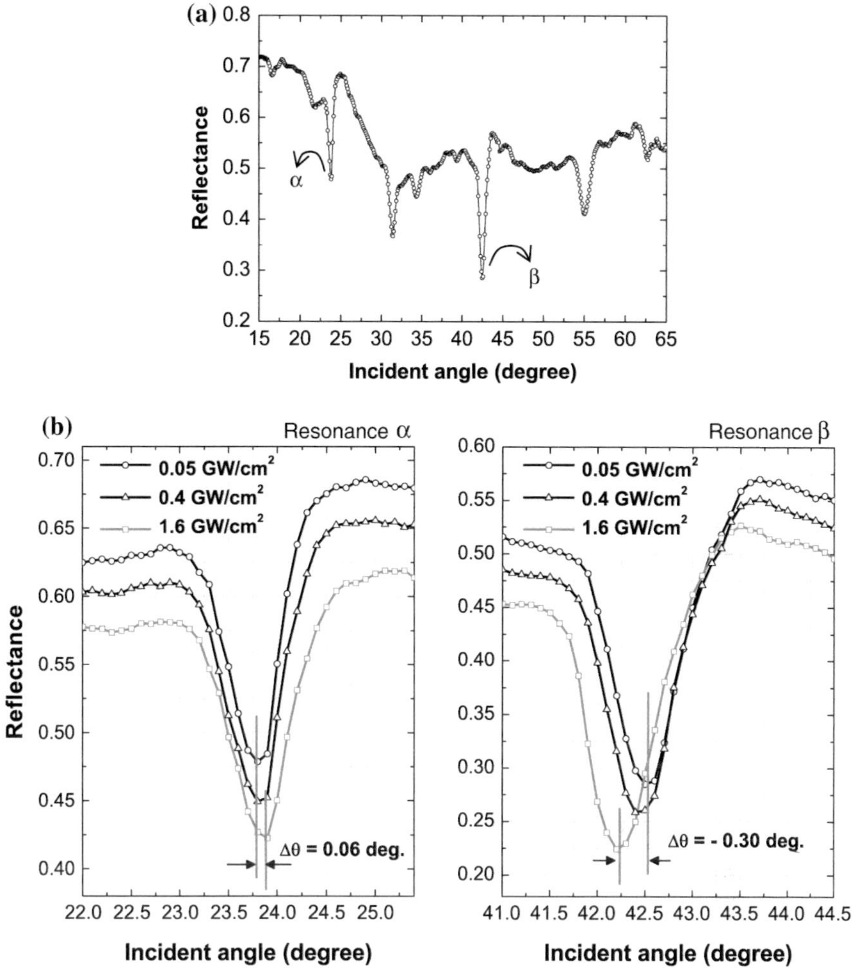

Fig. 2.9 Examples of linear and nonlinear angle-dependent reflectivities of a 2D photonic crystal waveguide with Kerr nonlinearity. **a** Linear angle-dependent reflectivity at a wavelength of 800 nm, measured with TE polarization along the $\Gamma - X$ lattice direction. **b** Nonlinear angle-dependent reflectivities for three different incident peak powers at two different angle ranges in the vicinity of the resonances α (*left*) and β (*right*), measured at a fixed wavelength of 800 nm along the $\Gamma - X$ direction. For clarity, the reflectivities are shifted by -0.03 along the vertical axis with respect to each other. Here, the vertically unshifted reflectivity used as a base is for a peak power of $0.05\,\mathrm{GW/cm^2}$ (*black circles*)

bands without having to consider the energy dispersions of the linear and nonlinear susceptibilities.

Figure 2.9b shows the angle-dependent reflectivities for different incident peak powers at a wavelength of 800 nm in two different angle ranges in the vicinity of the resonance angles α and β. Clearly, the resonance dip positions in the reflectivities are shifted with increasing incident peak power for both resonance dips. The angular shifts at resonances α and β have different sign. These shifts are accompanied by large nonlinear changes in the reflectivities at the resonance angles. These novel and unusual nonlinear phenomena, exhibiting differences in sign and quantity of the nonlinear angular shifts at the resonances, indicate that the nonlinear changes are strongly influenced by the photonic band dispersion properties.

To find the underlying physical mechanism of these results, Fig. 2.10a shows the experimental photonic band structure along the $\Gamma - X$ line of the sample examined by angle-dependent optical reflectivity measurements compared with the theoretical band structure calculated by the 3D-FDTD method for a model structure identical to the experimental sample structure (see scheme in Sect. 2.3). The features of the observed band dispersion are in reasonably good agreement with the theoretical curves. The experimental and theoretical band structures provide a very clear picture of the relation between the observed nonlinear responses and the photonic band dispersion. In Fig. 2.10a, the horizontal solid line shows the photon energy corresponding to the energy of the incident laser used in the nonlinear experiments. The positions of the in-plane wave vectors k of the incident light at the angles of resonances α and β at this photon energy are indexed as points I and II, respectively. The group velocities v_g of these corresponding bands, estimated by the slope of the dispersion curve ($v_g = \partial\omega/\partial k$), are $v_g = 0.46c$ and $v_g = -0.19c$ at points I and II, respectively.

We now directly compare the observed nonlinear optical responses with the photonic band structure features in the vicinity of the resonances. The k shift values obtained from the observed resonance angle shifts in Fig. 2.9b are $\Delta ka/2\pi = +9.5 \times 10^{-4}$ for resonance α and $\Delta ka/2\pi = -3.9 \times 10^{-3}$ for resonance β, where a is the lattice constant. The refractive index change Δn of the nonlinear optical polymer layer can be estimated by equating the observed Δk amplitude with the value of the refractive index change in the 3D-FDTD band calculations, giving $\Delta n = +1.7 \times 10^{-3}$ and $\Delta n = +3.7 \times 10^{-3}$ at α and β, respectively. The detailed linear band structure and nonlinear band shifts in the vicinity of I and II are shown in Fig. 2.10b. The shifts in the photon energy domain are towards the low photon energy side for both bands because of the positive susceptibility of the nonlinear optical polymer at the energy of the incident laser. On the other hand, the bands are shifted in opposite directions in the k domain, with a positive shift at α and a negative shift at β, which results from the positive and negative slopes of the band dispersion curves at these resonances. The agreement between the band structure features and the resonance angle shifts demonstrates the essential role of the band dispersion character in the nonlinear dynamics.

A direct comparison between the observed nonlinear responses and the dispersion characteristics for the two different band components provides information on the group velocity of the nonlinear optical properties and clarifies the enhancement effect of the nonlinear processes in photonic crystal systems. In the work, we found basic agreement between the ratio of the inverse group velocities of the photonic bands at I and II, $|v_{g_I}|^{-1} : |v_{g_{II}}|^{-1} = 1.0 : 2.4$, and the ratio of the refractive index changes α and β, $\Delta n\alpha : \Delta n\beta = 1.0 : 2.2$, as well as between the ratio of the squared inverse

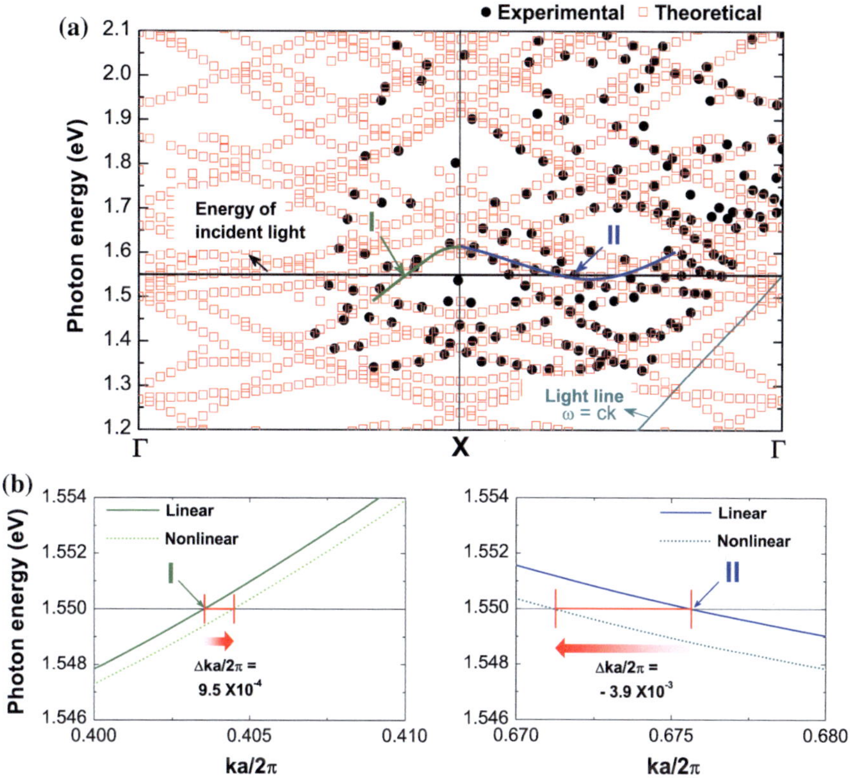

Fig. 2.10 **a** Experimental photonic band structure of the 2D PhC waveguide along the $\Gamma - X$ line obtained by polarized angular-dependent reflectivity measurements (*closed circles*), and theoretical band structure calculated by the 3D FDTD method (*open squares*). No empirical adjustments of any kind were made in the calculations. Points I and II are the positions of the wave vectors k and photon energies at the resonances α and β (see Fig. 2.9). The photonic bands contributing to the nonlinear reflectivity changes in the vicinity of α and β are indicated by *solid lines*. **b** Detailed theoretical linear band structure (*solid lines*) in the vicinity of points I (*left*) and II (*right*) and the nonlinear band shifts (*dashed lines*) obtained from an analysis of the observed resonance angle shifts in Fig. 2.9

group velocities at I and II, $|v_{\mathrm{gI}}|^{-2} : |v_{\mathrm{gII}}|^{-2} = 1.0 : 5.8$, and the ratio of the observed nonlinear resonance angle shifts at α and β, $|\Delta\theta\alpha| : |\Delta\theta\beta| = 1.0 : 5.0$. These results clearly demonstrate that the observed nonlinear changes are dominated by the band group velocities at the resonance conditions. The refractive index change in these processes is considered to be enhanced in proportion to v_{g}^{-1} because the electric field intensity of the resonance band mode is enhanced in proportion to v_{g}^{-1} based on conservation of energy in the first approximation. The resonance angle shift is enhanced in proportion to v_{g}^{-2} because the wave vector shift Δk, which depends on the resonance angle shift, is affected by the combination of the larger refractive index change ($\propto v_g^{-1}$), and the larger *k* vector shift ($\propto v_g^{-1}$) arising from the slower group velocity, as shown in Fig. 2.10b.

These data gives direct evidence that the nonlinear optical changes arising from modifications of the photonic bands by an external laser field resonantly coupled to the photonic band (i.e., by purely optical means) in a 2D photonic crystal waveguide with a Kerr nonlinearity are dominated by the dispersion nature and the group velocity of the photonic bands. Thus, with current technology, active manipulation of these nonlinear optical processes is a realistic possibility through engineering of the band dispersion and band group velocity characteristics.

2.4.2 Enhancement of Two-Photon Excited Fluorescence in Photonic Crystals

We want to conclude this chapter by presenting another category of nonlinear processes enhanced due to photonic crystal structures, namely that of two-photon excited fluorescence (TPEF). Two-photon excitation (TPE) processes have attracted considerable interest because of their potential to be applied to diverse fields, including three-dimensional fluorescence imaging [92, 93], high-density optical data storage [94], lithographic microfabrication [95], and photodynamic cancer therapy [96]. From the viewpoint of such applications, a third-order nonlinear TPE process has two important advantages over a single-photon excitation process: the ability to generate excited states using photons with half the linear excitation energy and a quadratic dependence on laser intensity. These advantages can increase the light penetration depth in absorbing media and provide a high three-dimensional spatial resolution under tight-focusing conditions, without being restrained by the diffraction limit. Unfortunately, the majority of known organic molecules (and most inorganic materials) have very small nonlinear absorption cross sections (σ), typically $\sigma \sim 1\,\mathrm{GM}(=10^{-50}\,\mathrm{cm}^4\,\mathrm{s\,photon}^{-1}\,\mathrm{molecule}^{-1})$. Consequently, very high laser powers are required, making the widespread use of TPEF impractical.

However, the combination of a photonic crystal structure and highly nonlinear materials can produce large nonlinearities, thus reducing the power requirements for TPE processes. The ability to control the light dispersion relation (i.e., the photonic band structure) including extremely slow group velocities c/v_g in photonic crystal systems enables nonlinear light-matter interactions to be enhanced due to the enhancement of the electromagnetic field of the excitation wave. In conventional waveguides, c/v_g is limited to the order of unity, whereas slower c/v_g in 2D photonic crystal waveguides can be achieved with many more degrees of freedom.

An example of a fabricated 2D photonic crystal slab waveguides produced from a highly nonlinear host polymer doped with molecules possessing a large two-photon absorption cross section are shown in Fig. 2.11a. A nonlinear optical molecule is the derivative of bis(styryl)benzene that exhibits a very large TPE cross section σ (as high as 900 GM) [97] due to a large delocalized π-electron system, and has very high

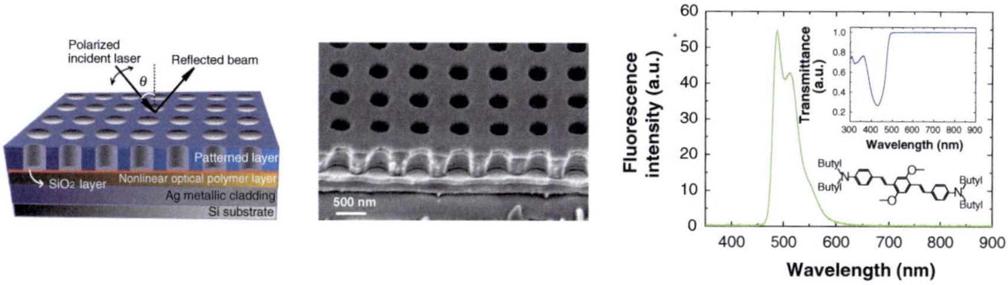

Fig. 2.11 **a** Schematic showing experimental geometry and SEM image showing cross section of fabricated nonlinear optical polymer 2D PhC waveguide. The patterned layer is a square lattice of 360-nm-diameter circular airholes with a lattice constant of 650 nm. **b** Linear fluorescence spectrum for the two-photon absorbing molecule used in the core of the 2D photonic crystal structure. The inset shows transmittance spectrum of the molecule and its molecular structure

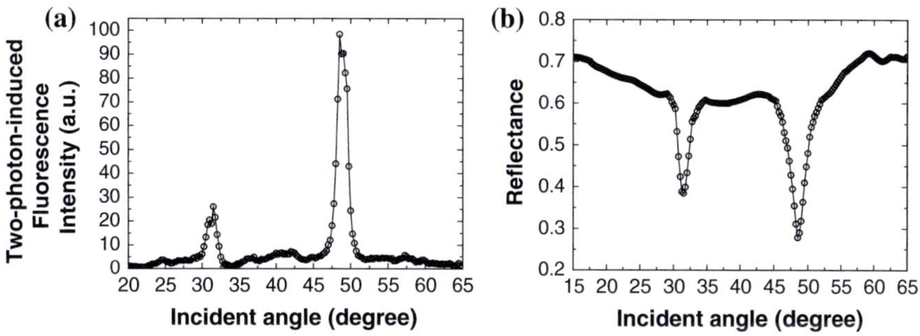

Fig. 2.12 **a** Incidence-angle dependence of TPEF intensity at 486 nm for the nonlinear optical polymer 2D photonic slab waveguide. **b** Angular dependence of the linear reflectivity of the same sample at a fixed wavelength of 800 nm along the $\Gamma - X$ direction

fluorescence quantum yields. Figure 2.11b shows the fluorescence and transmittance spectra of the core of the 2D photonic crystal structure; the molecular structure of the bis(styryl)benzene derivative is shown in the inset. For the specular TPEF measurements, a mode-locked Ti:sapphire laser operating at a wavelength of 800 nm is used as the excitation light source. To eliminate the pump wavelength beams, an interference filter that transmits only the TPEF beams is placed behind the sample. Further detailed techniques can be found in the Ref. [98].

Figure 2.12a shows the incident-angle dependence of the TPEF intensity at 486 nm for the highly nonlinear optical polymer 2D photonic crystal slab waveguide. The average pulse energy of the laser beam was 5 nJ. A strong peak was observed in the TPEF spectrum at an incidence angle of 47.5°. A large TPEF enhancement of approximately two orders of magnitude was achieved relative that for an unpatterned nonlinear optical polymer waveguide. In the low-incident-angle region, a weak peak was observed at 32.5° in the TPEF spectrum. These enhancements in the specular TPEF stem from resonant coupling between the fundamental external laser fields and the corresponding photonic band modes.

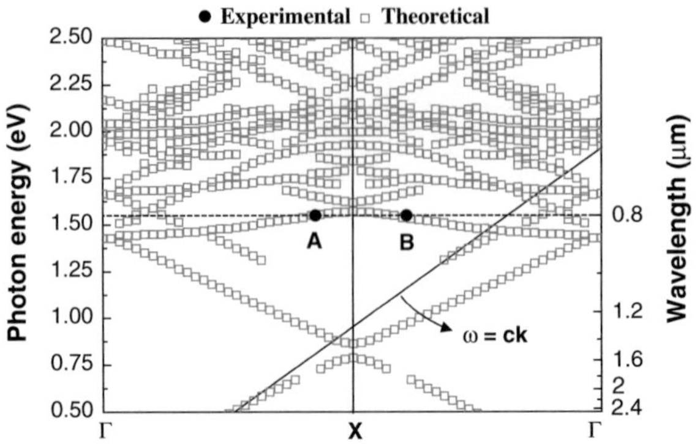

Fig. 2.13 Photonic band structure of the nonlinear optical polymer photonic crystal slab waveguide along the $\Gamma - X$ line calculated by the 3D-FDTD method (*open squares*). The experimentally probed positions (*closed circles*) of the resonance band modes in the polarized angular-resolved reflectivity are indexed as points A and B

To directly clarify the enhancement phenomena in the specular TPEF processes, linear-polarized angular-resolved reflectivity measurements were performed for the same sample. Figure 2.12b shows the reflectivity spectrum along the Γ-X direction for TE polarization at a wavelength of 800 nm. Two sharp dips (depressions) are clearly observed in the reflectivity. The observed angles of the dips are in good agreement with the TPEF enhancement angles at 32.5° and 47.5°. These dips originate from resonant phenomena associated with surface coupling to in-plane photonic bands at resonance energies and in-plane wave vectors. The observed resonance angles have associated in-plane wave vectors, $k = (2\pi/\lambda)\sin\theta$, where λ is the incident light wavelength and $\theta = 0°$ for normal incidence. The resonance dip at each resonance angle in the reflectivity is thus directly related to the in-plane photonic band mode. Accordingly, it was directly verified that the observed enhancements in the TPEF intensity originate from resonant couplings between the external pump laser field and the corresponding photonic band mode for the resonant energy and the in-plane wave vector.

Figure 2.13 shows the theoretical photonic band structure of the highly nonlinear optical polymer photonic crystal slab waveguide along the $\Gamma - X$ line calculated by the 3D-FDTD method for a model structure identical to the experimental sample structure. No empirical adjustments of any kind were made in the calculations. The black line corresponds to light in a vacuum ($\omega = ck$). The dashed line represents the photon energy corresponding to the energy of the incident laser used in the experiments. The experimentally probed positions of the resonance band modes in the polarized angular-resolved reflectivity are indicated by points A and B, and they agree well with the theoretical band structure. The experimental and theoretical band structures provide a useful picture of the relationship between the nonlinear optical responses and photonic band dispersions. The band dispersion branches denoted by points A and B correspond to the resonance band branches that contribute to the observed TPEF responses at incidence angles of 32.5° and 47.5° in Fig. 2.12a, respectively. Resonances A and B were found to be bands symmetric about the X point. The group velocities v_g of these bands were estimated to be $v_g = \pm 0.18c$ using the slope of the dispersion curve ($v_g = \partial\omega/\partial k$). This result demonstrates the enhancement of the TPEF resulting from the slower group velocity. By comparing the specular TPEF intensities of resonances A and B, it is reasonable to consider that these enhancements depend strongly on the resonance in-plane k vectors, that is, the surface coupling efficiency of the incident light fields.

The enhancement of the nonlinear optical processes resulting from the slower group velocity, such as demonstrated in this section, should lead to dramatic improvements in the performance of TPE and all-optical switching devices, and will open up new possibilities in nonlinear active applications. Furthermore, the agreement between the observed nonlinear responses and the band dispersion characteristics means we can begin to engineer the nonlinear optical properties of photonic crystal systems by controlling the photonic crystal configuration. We note that the features and advantages of photonic crystals are also common to other nonlinear optical processes, suggesting the possibility of progress in all areas of nonlinear optics.

References

1. J.W. Strutt (Lord Rayleigh), Phil. Mag. 24, 145 (1887)
2. J.W. Strutt (Lord Rayleigh), Phil. Mag. 26, 256 (1888)
3. C.G. Darwin, Phil. Mag. **27**, 675 (1914)
4. K. Ohtaka, Phys. Rev. B **19**, 5057 (1979)
5. E. Yablonovitch, Phys. Rev. Lett. **58**, 2059 (1987)
6. E. Yablonovitch, T.J. Gmitter, Phys. Rev. Lett. **63**, 1950 (1989)
7. S. John, Phys. Rev. Lett. **58**, 2486 (1987)
8. E. Yablonovitch, T.J. Gmitter, R.D. Meade, A.M. Rappe, K.D. Brommer, J.D. Joannopoulos, Phys. Rev. Lett. **67**, 3380 (1991)
9. K.M. Ho, C.T. Chan, C.M. Soukoulis, Phys. Rev. Lett. **65**, 3152 (1990)
10. P. Vukusic, J.R. Sambles, C.R. Lawrence, R.J. Wootton, Nature **410**, 36 (2001)
11. H. Ghiradella, Appl. Opt. **30**, 3492 (1991)
12. T.F. Anderson, A.G. Richards, J. Appl. Phys. **13**, 748 (1942)
13. A.R. Parker, P.C. Mcphedran, D.R. Mckenzie, L.C. Botten, N.P. Nicorovici, Nature **409**, 36 (2001)
14. J.W. Galusha, L.R. Richey, J.S. Gardner, J.N. Cha, M.H. Bartl, Phys. Rev. E 77, 050904(R) (2008).
15. J. Huang, X. Wang, Z.L. Wang, Nano Lett. **6**, 2325 (2006)
16. S.A. Davis, S.L. Burkett, N.H. Mendelson, S. Mann, Nature **385**, 420 (1997)
17. R. Seshadri, F. Meldrum, Adv. Mater. **12**, 1149 (2000)
18. O.D. Velev, T.A. Jede, R.F. Lobo, A.M. Lenhoff, Nature **389**, 447 (1997)
19. B.T. Holland, C.F. Blanford, A. Stein, Science **281**, 538 (1998)
20. R.C. Schroden, M. Al-Daous, C.F. Blanford, A. Stein, Chem. Mater. **14**, 3305 (2002)
21. O. Toader, S. John, K. Busch, Opt. Express **8**, 223 (2001)
22. S.Y. Lin, J.G. Fleming, D.L. Hetherington, B.K. Smith, R. Biswas, K.M. Ho, M.M. Sigalas, W. Zubrzycki, S.R. Kurtz, J. Bur, Nature **394**, 251 (1998)
23. S. Noda, K. Tomoda, N. Yamamoto, A. Chutinan, Science **289**, 604 (2000)

24. K. Aoki, H.T. Miyazaki, H. Hirayama, K. Inoshita, T. Baba, K. Sakoda, N. Shinya, Y. Aoyagi, Nat. Mater. **2**, 117 (2003)
25. M. Deubel, G.V. Freymann, M. Wegener, S. Pereira, K. Busch, C.M. Soukoulis, Nat. Mater. **3**, 444 (2004)
26. S.A. Rinne, F. Garcia-Santamaria, P.V. Braun, Nat. Photonics **2**, 52 (2007)
27. T.F. Krauss, R.M. De La Rue, S. Brand, Nature **383**, 699 (1996)
28. S.Y. Chou, P.R. Krauss, P.J. Renstrom, Appl. Phys. Lett. **67**, 3114 (1995)
29. S.Y. Chou, P.R. Krauss, P.J. Renstrom, Science **272**, 85 (1996)
30. M. Okinaka, S. Inoue, K. Tsukagoshi, Y. Aoyagi, J. Vac. Sci. Technol. B **24**, 271 (2006)
31. G.-Y. Jung, E. Johnston-Halperin, W. Wu, Z. Yu, S.-Y. Wang, W.M. Tong, Z. Li, J.E. Green, B.A. Sheriff, A. Boukai, Y. Bunimovich, J.R. Heath, R.S. Williams, Nano Lett. **6**, 351 (2006)
32. M. Okinaka, S. Inoue, K. Tsukagoshi, Y. Aoyagi, J. Vac. Sci. Technol. B **25**, 899 (2007)
33. M.D. Austin, H. Ge, W. Wu, M. Li, Z. Yu, D. Wasserman, S.A. Lyon, S.Y. Chou, Appl. Phys. Lett. **84**, 5299 (2004)
34. K.M. Leung, Y.F. Liu, Phys. Rev. Lett. **65**, 2646 (1990)
35. M. Plihal, A.A. Maradudin, Phys. Rev. B **44**, 8565 (1991)
36. J.D. Joannopoulos, R.B. Meade, J.N. Winn, *Photonic Crystals: Molding the Flow of Light* (Princeton University Press, Princeton, 1995)
37. K. Sakoda, Phys. Rev. B **55**, 15345 (1997)
38. S. Shi, C. Chen, D.W. Prather, J. Opt. Soc. Am. A **21**, 1769 (2004)
39. K. Sakoda, *Optical Properties of Photonic Crystals* (Springer, Berlin, 2004)
40. L. Shen, S. He, J. Opt. Soc. Am. A **19**, 1021 (2002)
41. A. David, H. Benisty, C. Weisbuch, Phys. Rev. B **73**, 075107 (2006)
42. K.S. Yee, IEEE Trans. Antennas Propag. **14**, 302 (1966)
43. A. Taflow, *Computational Electrodynamics: The Finite-Difference Time-Domain Method* (Artech House INC, Norwood, 1995)
44. K.S. Kunz, R.J. Luebbers, *The Finite Difference Time Domain Method for Elctromagnetics* (CRC Press, Boca Raton, 1993)
45. T. Weiland, Electronics and Communication AEUE **31**, 116 (1977)
46. R. Holland, IEEE Trans. Nucl. Sci. **30**, 4589 (1983)
47. W. Yu, R. Mittra, T. Su, Y. Liu, X. Yang, *Parallel Finite-Difference Time-Domain Method* (Artech House INC, Massachusetts, 2006)
48. W. Yu, X. Yang, Y. Liu, R. Mittra, *Electromagnetic Simulation Techniques Based on the FDTD Method* (Wiley, Hoboken, 2009)
49. G. Mur, IEEE Trans. Electromagn. Compat. **23**, 377 (1981)
50. Z. Yusheng, W. Wenbing, IEEE Microwave and Guided Wave Lett. **6**, 120 (1996)
51. J. Berenger, J. Comput. Phys. **114**, 185 (1994)
52. S.D. Gedny, IEEE Trans. Antennas Propag. **44**, 1630 (1996)
53. V.N. Astratov, D.M. Whittaker, I.S. Culshaw, R.M. Stevenson, M.S. Skolnick, T.F. Krauss, R.M. De La Rue, Phys. Rev. B **60**, R16255 (1999)
54. D.M. Whittaker, I.S. Culshaw, Phys. Rev. B **60**, 2610 (1999)
55. S. Inoue, Y. Aoyagi, Phys. Rev. B **69**, 205109 (2004)
56. M.S. Thijssen, R. Sprik, J.E.G.J. Wijnhoven, M. Megens, T. Narayanan, Phys. Rev. Lett. **83**, 2730 (1999)
57. M. Notomi, K. Yamada, A. Shinya, J. Takahashi, C. Takahashi, I. Yokohama, Phys. Rev. Lett. **87**, 253902 (2001)
58. A. Otto, Z. Phys. **216**, 398–410 (1968)
59. S. Inoue, S. Yokoyama, Y. Aoyagi, Opt. Express **16**, 2461 (2008)
60. S.Y. Lin, E. Chow, V. Hietala, P.R. Villeneuve, J.D. Joannopoulos, Science **282**, 274 (1998)
61. S.J. McNab, N. Moll, Y.A. Vlasov, Opt. Express **11**, 2927 (2003)
62. P.I. Borel, A. Harpoth, L.H. Frandsen, M. Kristensen, P. Shi, J.S. Jensen, O. Sigmund, Opt. Express **12**, 1996 (2004)
63. P. Lodahl, A. Floris van Driel, I.S. Nikolaev, A. Irman, K. Overgaag, D. Vanmaekelbergh, W.L. Vos, Nature **430**, 654 (2004)

64. D. Englund, D. Fattal, E. Waks, G. Solomon, B. Zhang, T. Nakaoka, Y. Arakawa, Y. Yamamoto, J. Vuckovic, Phys. Rev. Lett. **96**, 013904 (2005)
65. S. Noda, M. Fujita, T. Asano, Nat. Photonics **1**, 449 (2007)
66. Y. Akahane, T. Asano, B. Song, S. Noda, Nature **425**, 944 (2003)
67. T. Yoshie, A. Scherer, J. Hendrickson, G. Khitrova, H.M. Gibbs, G. Rupper, C. Ell, O.B. Shchekin, D.G. Deppe, Nature **432**, 200 (2004)
68. S.L. Portalupi, M. Galli, C. Reardon, T.F. Krauss, L. O'Faolain, L.C. Andreani, D. Gerace, Opt. Express **15**, 16064 (2010)
69. E. Cubukcu, K. Aydin, E. Ozbay, S. Foteinopoulou, C.M. Soukoulis, Nature **423**, 604 (2003)
70. S. Foteinopoulou, C.M. Soukoulis, Phys. Rev. B **67**, 235107 (2003)
71. A. Berrier, M. Mulot, M. Swillo, M. Qiu, L. Thylen, A. Talneau, S. Anand, Phys. Rev. Lett. **93**, 073902 (2004)
72. S. Kocaman, R. Chatterjee, N.C. Panoiu, J.F. McMillan, M.B. Yu, R.M. Osgood, D.L. Kwong, C.W. Wong, Phys. Rev. Lett. **102**, 203905 (2009)
73. P. Russell, Science **299**, 358 (2003)
74. J.C. Knight, Nature **424**, 847 (2003)
75. T. Baba, Nat. Photonics **2**, 465 (2008)
76. J.J. Wierer, A. David, M.M. Megens, Nat. Photonics **3**, 163 (2009)
77. Y. Chassagneux, R. Colombelli, W. Maineult, S. Barbieri, H.E. Beere, D.A. Ritchie, S.P. Khanna, E.H. Linfield, A.G. Davies, Nature **457**, 174 (2009)
78. H. Kosaka, T. Kawashima, A. Tomita, M. Notomi, M. Notomi, T. Tamamura, T. Sato, S. Kawakami, Phys. Rev. B **58**, R10096 (1998)
79. K. Guven, K. Aydin, K.B. Alici, C.M. Soukoulis, E. Ozbay, Phys. Rev. B **58**, 205125 (2004)
80. M. Scalora, J.P. Dowling, C.M. Bowden, M.J. Bloemer, Phys. Rev. Lett. **73**, 1368 (1994)
81. M. Soljacis, J.D. Joannopoulos, Nat. Mater. **3**, 211 (2004)
82. S. Inoue, Y. Aoyagi, Phys. Rev. Lett. **94**, 103904 (2005)
83. X. Hu, P. Jiang, C. Ding, H. Yang, Q. Gong, Nat. Photonics **2**, 185 (2008)
84. K. Sakoda, K. Ohtaka, Phys. Rev. B **54**, 5742 (1996)
85. A.R. Cowan, J.F. Young, Phys. Rev. B **65**, 5751 (2002)
86. S. Inoue, Y. Aoyagi, Jpn. J. Appl. Phys. **45**, 6103 (2006)
87. S. Inoue, S. Yokoyama, Thin Solid Films **518**, 470 (2009)
88. K. Rivoire, S. Buckley, J. Vuckovi, Opt. Express **19**, 22198 (2011)
89. C. Monat, B. Corcoran, M. Ebnali-Heidari, C. Grillet, B.J. Eggleton, T.P. White, L. O'Faolain, T.F. Krauss, Opt. Express **17**, 2944 (2009)
90. J.M. Dudley, J.R. Taylor, Nat. Photonics **3**, 85 (2009)
91. R.E. Slusher, B.J. Eggleton, *Nonlinear Photonic Crystals* (Springer, Berlin, 2003)
92. W. Denk, J.H. Strickler, W.W. Webb, Science **248**, 73 (1990)
93. R.H. Kohler, J. Cao, W.R. Zipfe, W.W. Webb, M.R. Hansen, Science **276**, 2039 (1997)
94. D.A. Parthenopoulos, P.M. Rentzepis, Science **245**, 843 (1989)
95. B.H. Cumpston, S.P. Ananthavel, S. Barlow, D.L. Dyer, J.E. Ehrlich, L.L. Erskine, A.A. Heikal, S.M. Kuebler, I.-Y.S. Lee, D. McCord-Maughon, J. Qin, H. Rockel, M. Rumi, X.-L. Wu, S.R. Marder, J.W. Perry, Nature **398**, 51 (1999)
96. J.D. Bhawalkar, G.S. He, P.N. Prasad, Rep. Prog. Phys. **59**, 1041 (1996)
97. M. Kuebler, I.-Y.S. Lee, D. McCord-Maughon, J. Qin, H. Rockel, M. Rumi, X.-L. Wu, S.R. Marder, J.W. Perry, Nature **398**, 51 (1999)
98. S. Inoue, S. Yokoyama, Appl. Phys. Lett. **93**, 111110 (2008)

Chapter 3
Surface Plasmons

Kotaro Kajikawa

Abstract The surface plasmon is collective oscillation of free electrons localized at a metallic surface. Metals had not been extremely exciting optical materials for long time. However, many research works on surface plasmons have been reported in the last decade, since surface plasmons have attractive optical properties that cannot be realized in dielectrics and semiconductors. This chapter briefly deals with the basic theory and some applications of surface plasmons.

3.1 General Remarks

The surface plasmon is collective oscillation of free electrons localized at a metallic surface. It has the following attractive optical characteristics. (a) The electric field of light is largely enhanced. This enables us to make highly sensitive optical measurements and spectroscopy. (b) The resonance condition is sensitive to the permittivity around the metallic structure. Highly sensitive optical sensors are developed with this feature. (c) The light energy is confined in a nanometer-sized region as surface plasmons. It has been demonstrated that thin metallic wires work as a nanometer-sized plasmonic waveguide. These features will enable us to develop new types of optical devices.

The surface plasmons are roughly categorized into two. One is the propagating surface plasmon, sometimes called the planner surface plasmon, and the other is the localized surface plasmon (LSP). The former is surface plasmons propagating along a metallic surface. It is widely used in sensing of biological molecules such as DNA and proteins. The latter is surface plasmons excited in metallic nanostructures. It is widely used as a sensitizing methods for spectroscopy, such as surface enhanced

K. Kajikawa (✉)
Interdisciplinary Graduate School of Science and Engineering, Tokyo Institute of Technology, Nagatsuta, Midori-ku, Yokohama, 226-8502 , Japan
e-mail: kajikawa@ep.titech.ac.jp

Y. Aoyagi and K. Kajikawa (eds.), *Optical Properties of Advanced Materials*,
Springer Series in Materials Science 168, DOI: 10.1007/978-3-642-33527-3_3,

Raman scattering spectroscopy and surface enhanced infrared spectroscopy, and is also used as the sensing of biological molecules such as DNA and proteins. This section briefly introduces basic features of surface plasmons and their applications.

3.2 Propagating Surface Plasmons

In general, bulk plasmons are not coupled with light since they are longitudinal wave modes whereas the light wave propagating in a free space is transverse. Suppose that light having wave number k is incident to an interface between medium 1 (permittivity ϵ_1) and medium 2 (permittivity ϵ_2), as shown in Fig. 3.1. The wavenumber k is

$$k = \frac{2\pi}{\lambda}, \tag{3.1}$$

where λ is the wavelength of light. The bulk plasma waves are excited in the metal. While the bulk plasma waves are longitudinal, there exists a transverse component adjacent to the surface and interface. This mode is called surface plasma or surface plasmon, which can be coupled with light.

Considering the boundary condition at the surface, the normal component of the wavevectors satisfies the following relation,

$$\frac{k_{1z}}{\epsilon_1} + \frac{k_{2z}}{\epsilon_2} = 0, \tag{3.2}$$

according to the Maxwell's equations. Here k_{iz} is the normal component of wavevector in medium i.

The Snell's low, $k_{1x} = k_{2x} = k_x$, gives the relation

$$k_{zi}^2 = \sqrt{\epsilon_1}\left(\frac{\omega}{c}\right)^2 - k_x^2. \tag{3.3}$$

By using this relation, we can solve the Eq. 3.2.

$$k_x = \frac{\omega}{c}\sqrt{\frac{\epsilon_1\epsilon_2}{\epsilon_1 + \epsilon_2}}, \tag{3.4}$$

which is the dispersion relation of surface plasmons. This relation is schematically shown in Fig. 3.2, [1–3] together with the dispersion relation of light propagating in a free space (called light line) with the solid line. Since there is no point of intersection between the dispersion relations, the surface plasmons are not excited by the light. If we have an evanescent light having the dispersion relation shown with the broken line, the surface plasmons are excited at $k = k_{\text{sp}}$. The evanescent light is usually generated when light is total-reflected using a prism or a waveguide.

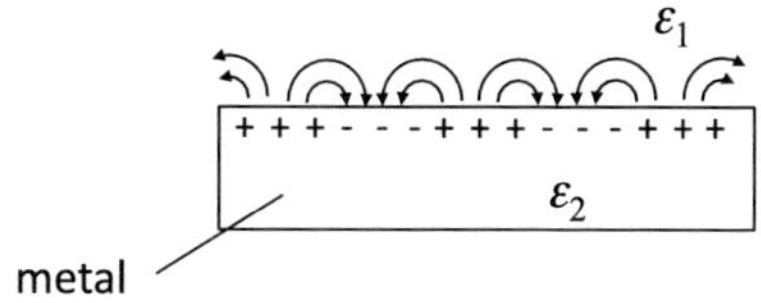

Fig. 3.1 Schematic picture of propagating surface plasmons

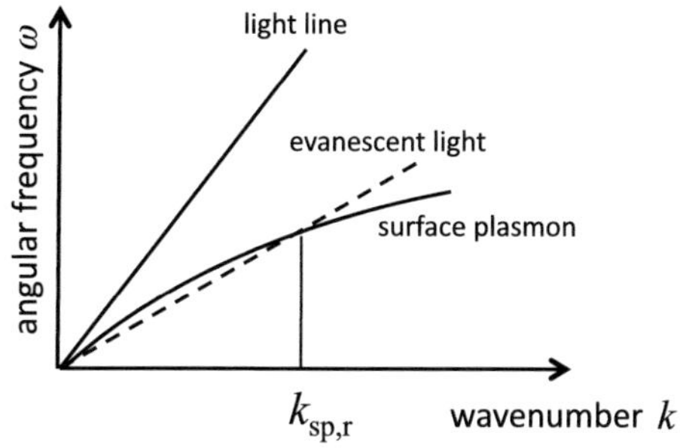

Fig. 3.2 The dispersion relation

Using the Eq. 3.3, the complex wave number $k = k' + k''i$ is written using the complex permittivity $\epsilon_2 = \epsilon_2' + \epsilon_2''i$ as,

$$k_x' = \left(\frac{\omega}{c}\right)\sqrt{\epsilon_1\epsilon_2'/\epsilon_1 + \epsilon_2'} \tag{3.5}$$

$$k_x'' = \left(\frac{\omega}{c}\right)(\epsilon_1\epsilon_2'/\epsilon_1 + \epsilon_2')^{\frac{3}{2}}(\epsilon_2''/2(\epsilon_2')^2) \tag{3.6}$$

According to these relations, the propagation length of the surface plasmon L_p is

$$L_\mathrm{p} = \frac{1}{2}k_x''. \tag{3.7}$$

These relations are plotted for gold and silver in Fig. 3.3. For the light with a wavelength 700–500 nm, the propagation lengths of surface plasmons are evaluated to be 20–30 μm (silver) and 5 μm (gold) [4].

For exciting surface plasmons, it is needed to match the momentums of light and surface plasmons. The attenuated total reflection (ATR) geometry is widely used to perform the phase matching. Two kinds of configurations are proposed: one the Kretschmann configuration (Fig. 3.4a) [5] and the other the Otto configuration (Fig. 3.5a) [6]. In the Kretschmann configuration, a thin metallic film with a thickness of approximately 50 nm is deposited at the bottom of an ATR prism. The solid line of Fig. 3.4b shows a calculated reflectivity profile at 633 nm for an ATR prism of BK7 ($n = 1.52$). There is a sharp dip at an angle of incidence $\theta_\mathrm{r} = 44°$ ascribed to the surface plasmon resonance. A largely enhanced electric field is generated in adjacent to the surface of the metallic thin film. This comes from the large wavenumber of the surface plasmon propagating along the metallic surface. The resonance angle

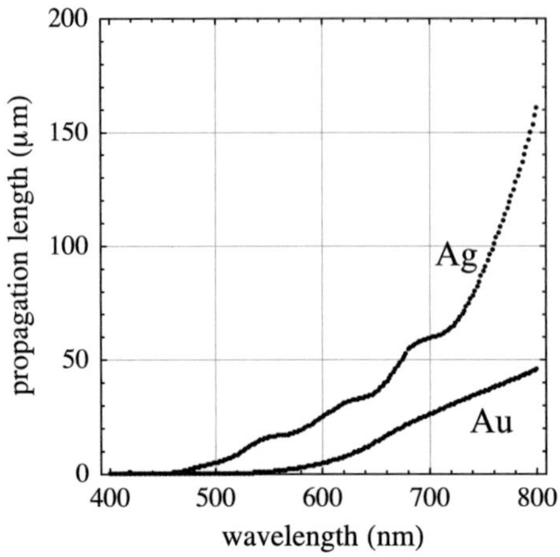

Fig. 3.3 Propagation length

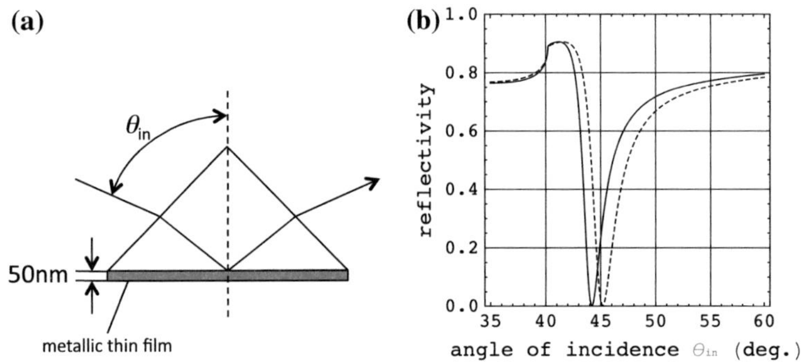

Fig. 3.4 Kertschmann configuration (**a**) and reflectivity profile (**b**)

is changed by the existence of dielectrics at the metallic surface or the refractive index of the ambient medium. The broken line of Fig. 3.4b shows the reflectivity profile when a 5 nm thick dielectric layer ($n = 1.5$) exists on the metallic surface. The resonance angle is by approximately 1° shifted to higher. Similar peak shift of 1° is calculated in the absence of the dielectric layer, when the refractive index of the ambient medium slightly increases ($n = 1.011$). Since we can determine the angle of incidence with an accuracy of 0.01–0.001°, the refractive index resolution is evaluated to be 10^{-5}. Since we can expose the metallic surface to ambient of air, water or liquid, using the Kretschmann configuration, it can be used for various sensing applications [7–10].

In the Otto configuration, a metallic surface is separated with a gap distance of a few hundreds of nanometers from the bottom of the prism. The gap is made of air or dielectrics of which refractive index is smaller than that of the ATR prism. Figure 3.5b shows the refractivity profiles at various gap distances. It is observed

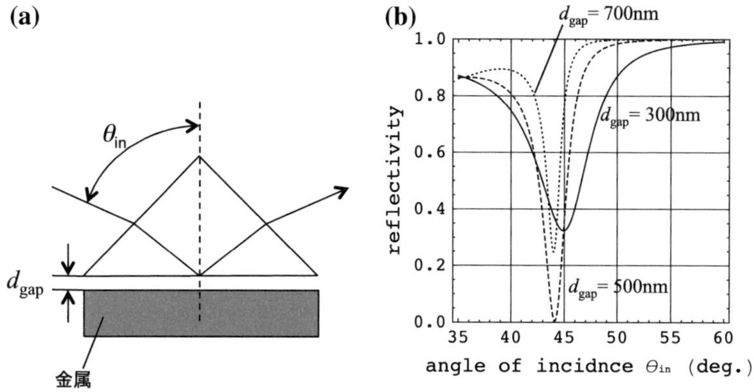

Fig. 3.5 Otto configuration (**a**) and reflectivity profile (**b**)

that the profile varies with the gap distance. The advantage of the Otto configuration over the Kretschmann is that surface plasmons can be excited in a bulk metal. Thus we can use this configuration for probing a bulk metallic surface. However, in practice, it is difficult to form the gap of hundreds nanometers. Hence the number of studies using the Otto configuration is small.

3.3 Localized Surface Plasmons

Localized surface plasmons (LSPs) occur in metallic nanostructures such as metallic rough surfaces and colloidal metals. It provides interesting optical properties that cannot be realized in other materials. Applications of localized surface plasmons involve surface enhanced Raman scattering, [11] biosensors, near-field optical microscopy [12] and nanometer optical waveguides [13]. Here we show a few systems of nanometer metallic particles that show localized surface plasmons.

3.3.1 LSPs in Nanospheres

A spherical colloidal gold solution is red, since the resonance wavelength of the localized surface plasmons in the colloidal gold is 510–520 nm. At the resonance wavelength, the colloidal gold has a maximum polarizability. Here we give the theoretical description of polarizability α. Suppose that a nanosphere with a permittivity ϵ_1 is located in an ambient medium m (permittivity ϵ_m). The radius of the nanosphere is a. The geometry is given in Fig. 3.6a. Then the polarizability of the nanosphere is written as [14],

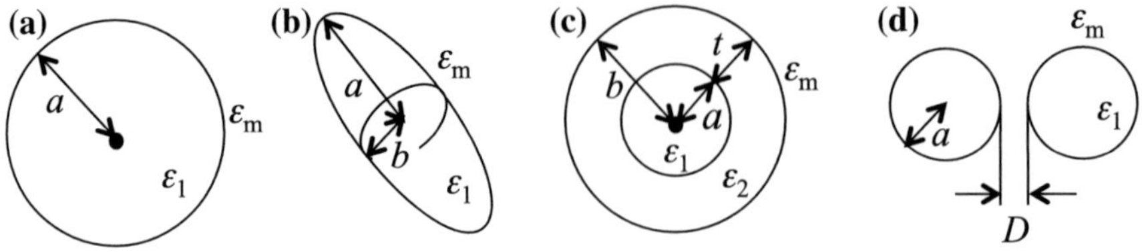

Fig. 3.6 **a** Nanosphere, **b** nanorod, **c** core-shell particle and **d** bisphere

$$\alpha = 4\pi\epsilon_m a^3 \frac{\epsilon_1 - \epsilon_m}{\epsilon_1 + 2\epsilon_m}. \tag{3.8}$$

The resonance condition occurs when the denominator is minimum. Since the permittivity of a metal depends on the wavelength of light, the polarizability α is a function of the wavelength of light and has a peak. The absorption cross section C_{abs} and scattering cross section C_{sca} are

$$C_{abs} = k\text{Im}\alpha \tag{3.9}$$

and

$$C_{sca} = \frac{k^4}{6\pi}|\alpha^2|, \tag{3.10}$$

where k is wavenumber. k is

$$k = \frac{2\pi}{\lambda}, \tag{3.11}$$

where λ is the wavelength of light. The absorption cross sections of gold or silver nanospheres C_{abs} are plotted as a function of the wavelength of light in Fig. 3.7. The dielectric constants of the metals are from literature [15]. The absorption is maximum at 350 nm for silver nanospheres and at 520 nm for gold nanosheres in air ($\epsilon_m = 1$). The peak wavelength is changed when the permittivity of ambient medium ϵ_m is different. Since the peak wavelength shift is sensitive to ϵ_m, it can be used as a transducer of sensors, such as refractive index, chemical and biological molecules.

3.3.2 LSPs in Nanorods

Another example is a colloidal particle of rotational ellipsoid that has long axes a and short axis b. The geometry is shown in Fig. 3.6b. The nanorod is composed of a medium with a permittivity ϵ_1, and is in an ambient medium ϵ_m. Two polarizability components exist for long axis α_a and for short one α_b. They are written as

$$\alpha_i = 4\pi ab^2\epsilon_m \frac{\epsilon_1 - \epsilon_m}{3\epsilon_m + 3L_i(\epsilon_1 - \epsilon_m)} \tag{3.12}$$

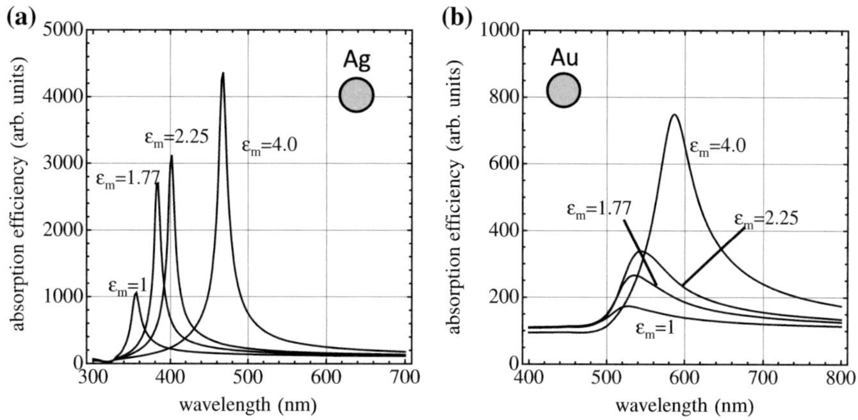

Fig. 3.7 Absorption efficiency of (**a**) silver and (**b**) gold nanospheres located in ambient of different ϵ_m

where $i = a$ or b. L_i's are depolarization factors for i direction [14]. When $a > b$, L_i's are

$$L_a = \frac{1-e^2}{e^2}\left(\frac{1}{2e}\log\frac{1+e}{1-e} - 1\right) \tag{3.13}$$

and

$$L_b = \frac{1-L_a}{2}. \tag{3.14}$$

The eccentricity e is defined as

$$e^2 = 1 - \frac{b^2}{a^2}. \tag{3.15}$$

When $a < b$, they are

$$L_a = \frac{1}{2e^2}\sqrt{\frac{1-e^2}{e^2}}\left(\frac{\pi}{2} - \tan^{-1}\sqrt{\frac{1-e^2}{e^2}}\right) - \frac{1-e^2}{2e^2} \tag{3.16}$$

and

$$L_b = \frac{1-L_a}{2} \quad \text{and} \quad e^2 = 1 - \frac{a^2}{b^2}. \tag{3.17}$$

The absorption efficiency of silver or gold nanoparticles is plotted as a function of the wavelength of light in Fig. 3.8, when $a > b$. In both nanorods, the resonance wavelength shift to the longer wavelength with increasing the aspect ratio a/b. When $a/b \geq 4$, the resonance wavelength lies in the near infrared region. Thus it can be

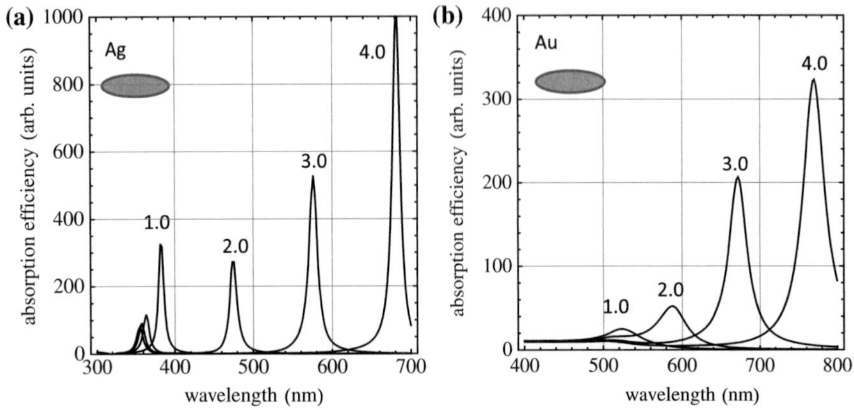

Fig. 3.8 Absorption efficiency of (**a**) silver and (**b**) gold nanorods in air at various aspect ratios

used for biological imaging [16, 17] since the near infrared is relatively transparent in biological tissues [18, 19].

3.3.3 LSPs in Core-Shell Spheres

Here we consider spheres coated with a thin layer. The geometry is shown in Fig. 3.6c. A core sphere (permittivity ϵ_a) with a radius a is covered with a thin layer (permittivity ϵ_b) with thickness t. The total radius is b in which $b = a + t$. The ambient permittivity is ϵ_m. Then, the polarizability α can be described as [20, 21]:

$$\alpha = 4\pi\epsilon_m b^3 \frac{\epsilon_2\epsilon_A - \epsilon_m\epsilon_B}{\epsilon_2\epsilon_A + 2\epsilon_m\epsilon_B} \tag{3.18}$$

where

$$\epsilon_A = \epsilon_1 + 2\epsilon_2 + 2(\epsilon_1 - \epsilon_2)\left(\frac{a}{b}\right)^3 \tag{3.19}$$

and

$$\epsilon_B = \epsilon_1 + 2\epsilon_2 - (\epsilon_1 - \epsilon_2)\left(\frac{a}{b}\right)^3 . \tag{3.20}$$

Here we consider two kinds of core-shell particles: particle A is composed of a metallic core covered with a SiO_2 dielectric shell (permittivity 2.25), and particle B is composed of a SiO_2 dielectric core covered with a metallic shell. Figure 3.9 shows the calculated absorption efficiency as a function of wavelength at various ratios t/a, where the size of the particle $(t + a)$ is constant. Four different core-shell nanoparticles are calculated: (a) core is silver and shell is SiO_2; (b) core is gold and shell is SiO_2; (c) core is SiO_2 and shell is silver; (d) core is SiO_2 and shell is gold.

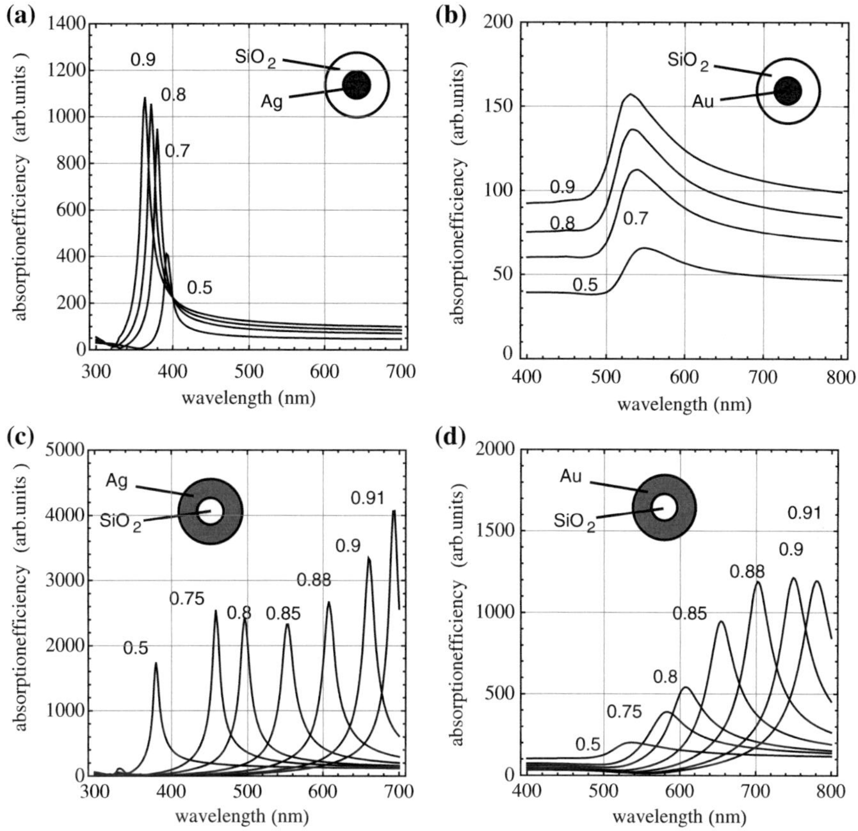

Fig. 3.9 Absorption efficiency of particles A (**a**) silver and (**b**) gold; particle B (**c**) silver and (**d**) gold, in air, at various a/b ratios

In particles A, the red-shift is observed with increasing the thickness of the shell (decreasing a/b), while the amount of the red-shift is not so large (less than 30 nm). In particles B, the red-shift is observed with decreasing the thickness of the shell (increasing a/b). The amount of the red-shift is very large. By changing the ratio a/b, the particles B can cover the spectral range from visible to the near infrared. Therefore, particle B is useful for applications of LSPs [21, 22].

3.3.4 Bispheres

The bisphere is an interesting particle that shows variety of spectrum. Figure 3.6d shows the geometry. Two spheres electromagnetically interacts each other and the resonance shifts to longer wavelengths. In addition, strong anisotropy appears. The amount of shift is a function of $D/(2a)$, where D is the separation and a is the radius of the particle. There are several methods proposed to calculate the optical response. Aravind et al. [23] and Ruppin [24] uses the bispherical coordinates to solve the Laplace equation when light electric field is induced. Wind et al. [25] solved this problem by use of the method of the multipole expansion in the spherical coordinates. Since the mathematical treatment is complicated, we only show here the calculated results. Figure 3.10a, b shows the absorption efficiency of the silver bispheres for two different polarization of incident light: (a) is the polarization along the long axis of the bisphere and (b) is the polarization normal to the long axis. By decreasing the ratio of $D/(2a)$, the absorption peak for parallel polarization is redshifted from 350 to 550 nm, whereas the absorption peak for normal polarization is scarcely moved. Namely, when the spheres approach, the absorption band is shifted to the longer wavelength, because of the strong interaction between the particles. Similar behavior is observed for gold bispheres as shown in Fig. 3.10c, d. In case of gold, the absorption band is rather broader. This is due to the large imaginary part of dielectric constants over the wavelengths, compared with those of silver.

3.4 Biosensors

3.4.1 Introduction

The biosensor is one of the most important applications of the surface plasmon resonance. It is a sensor that probes biological molecules, such as proteins, DNAs, and pathogens. This is a method to utilize the feature of surface plasmons that the resonance condition is sensitive to the presence of matters adjacent to the surface. It usually works without labeling of fluorescent molecules. Therefore it is a universal method to probe biological molecules. The principle is simple as illustrated in Fig. 3.11. The ligand that has affinity to the analyte is immobilized on the surface plasmon surface. Optically, ligand can be regarded as a dielectric mono- or submonomolecular layer on the surface, even it is not a perfect layer. When the analyte exists in a solution, the analyte binds to the ligand. For example, DNAs are immobilized on a metallic surface, the DNAs with complimentary base sequence can be detected. The binding of the analyte can be regarded as an increase in the thickness of surface layer. If one can detect the increase in the thickness using any methods, the methods can be used as a biological sensor. A good example is the quartz microbalance that can probe the increase of the surface layer as a change in the oscillation frequency [26, 27]. Optical methods are possible, but it is difficult to probe the change using a simple reflectivity measurement because the change in the reflectivity is very small in the order of 10^{-6}. Surface plasmon resonance is an effective probe for the detection of the surface layer because the change in the reflectivity or change in the scattering efficiency is typically in the order of 0.1. Now the surface plasmon resonance biosensors are widely used in various fields such as biochemistry, genome science and medicals.

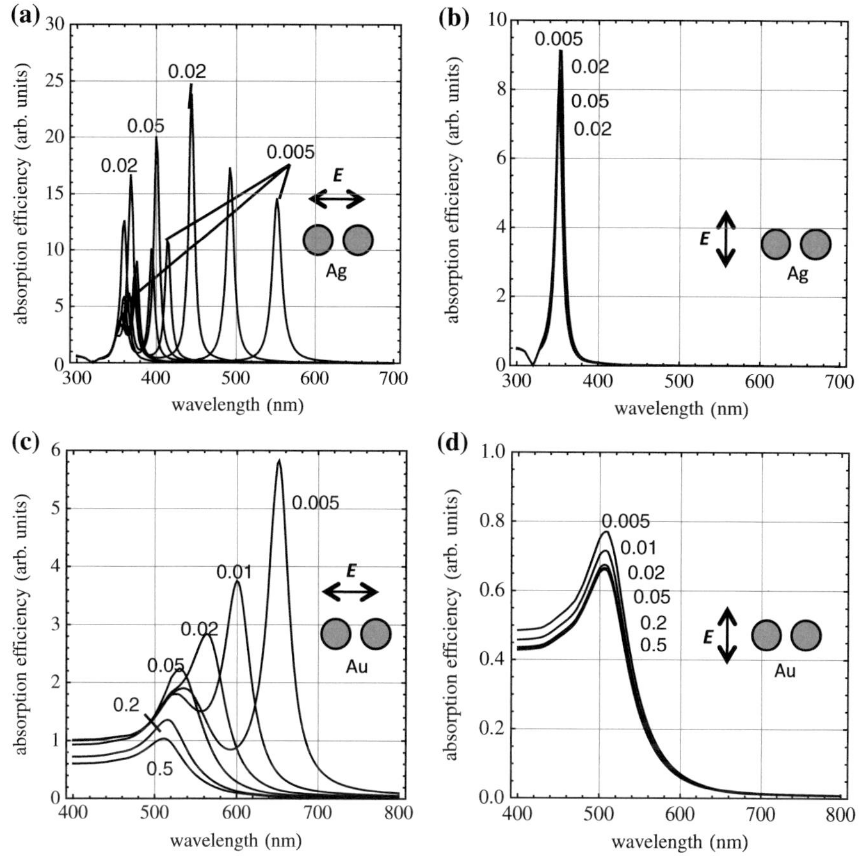

Fig. 3.10 Absorption efficiency of bispheres at different $D/(2a)$ ratios: (**a**) silver bisphere excited by parallel field to the long axis, (**b**) silver bisphere excited by perpendicular field, (**c**) gold bisphere excited by parallel field, and (**d**) gold bisphere excited by perpendicular field, in air

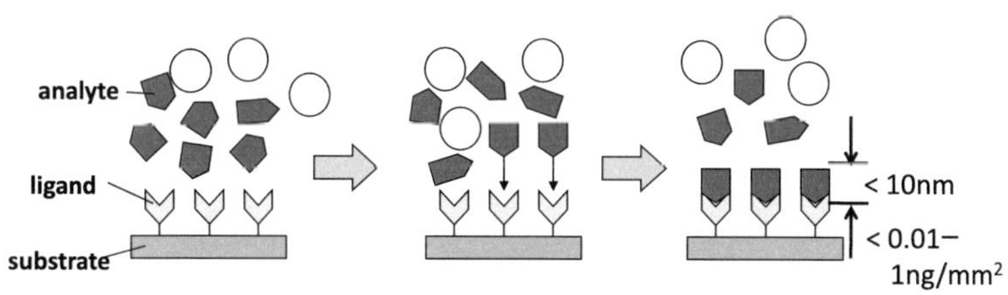

Fig. 3.11 Schematic picture of biosensing

3.4.2 Propagating Surface Plasmon Biosensors

Propagating surface plasmon biosensors using the Kretschmann configuration are the most widely used [7–10]. Figure 3.12a shows the geometry. A thin gold film 50 nm thick is deposited at the bottom of the prism. At the surface of the thin gold film, the ligand is immobilized. p-polarized light is incident at oblique incidence and the reflectivity is measured. With increasing the angle of incidence, attenuated total reflection (ATR) condition is achieved, and at a resonance angle, a sharp dip of reflectivity is observed. When analyte molecules bind to the ligand, the resonance angle is shifted to higher angles. Typically the shift of the resonance angle is 0.1–0.15° when analyte of 1 ng/mm^2 binds to the ligand. It seems that the measurements must be made by the mechanical change of the angle of incidence with a rotation stage. However such a mechanical system with a rotation stage is not always necessary. As shown in Fig. 3.12b, the incident light is expanded by use of a beam expander and focus the light on the surface with a lens. Then the indent light involves a wide range of incident angles. Detection of the reflected light with a photodiode array or a CCD camera allows us to measure the reflectivity profile at once, without using any mechanical apparatus. Kinetics of the binding events can be probed with this method since the detection time is usually in the order of 0.1–1 s. The features of this mode are (a) the system is mechanical-free and it is easy to use the apparatus. (b) Relatively high tolerance for the thickness of the gold film, and (c) high speed computer and

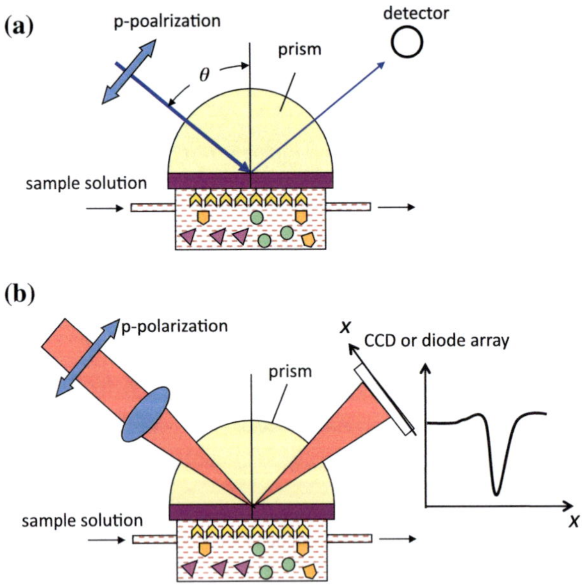

Fig. 3.12 Schematic picture of the surface plasmon resonance biosensor (**a**) changing the angle of incidence by mechanical rotation and (**b**) photodiode array or CCD camera detection

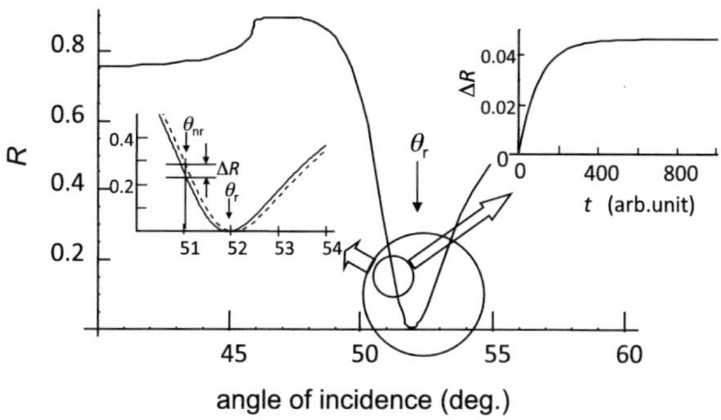

Fig. 3.13 The reflectivity as a function of the angle of incidence and kinetic measurement

algorithm are necessary to make quick evaluation of the resonance angle. Most of the commercially available biosensing apparatus adopt this mode.

Although the measurement of the resonance angle is the most usual way to measure the amount of the analyte molecules, reflectivity measurements at a fixed near-resonance angle also provides the kinetics of binding events. The scheme is depicted in Fig. 3.13. On binding of the analyte to ligand, the reflectivity profile shifts to the higher angle. This is indicated in the inset of the left-hand side that the solid line is changed to the broken line. When the angle of incidence is fixed to the near resonance angle of 51°, the reflectivity gradually increases about 0.04 as shown in the inset of right-hand side. The features of this mode are (a) alignment of the optical setup is necessary for the measurements with high accuracy, (b) the sensitivity is high but it depends on the angle of incidence. Hence the ATR scan is necessary beforehand to evaluate the amount of the analyte. (c) There are a few causes of the increase in reflectivity, other than the shift of the reflectivity profile: the broadening of the reflectivity profile upon inhomogenous binding of the analyte and the fluctuation of the incident light intensity. Therefore careful analysis is necessary to interpret the results. (d) A narrow tolerance for the thickness of the gold film since the sensitivity depends on the thickness. (c) It is relatively easy to make home-made biosensing apparatus using this mode. (f) The mass resolution is about an order of magnitude greater than the resonance angle mode.

There are some methods of propagating surface plasmon resonance without using any ATR prism. Instead of the ATR prism, optical waveguide can be used for generating the evanescent wave. Of various waveguides, optical fiber is promising because it is small and easy to handle. In addition coupling to the optical light source and detector is easy. Figure 3.14 shows an optical fiber biosensor using propagation surface plasmon resonance [28]. Since the core has higher refractive index, light propagates inside the core with total reflection. Evanescent wave is produced on total reflection. Then the clad is removed and gold thin film is deposited around the core. If one incident a white light, surface plasmon resonance occurs at a certain wavelength and the light at the resonance frequency is adsorbed by the gold film. Then a dip at the resonance frequency is observed in the transmittance spectrum. A ligand is deposited

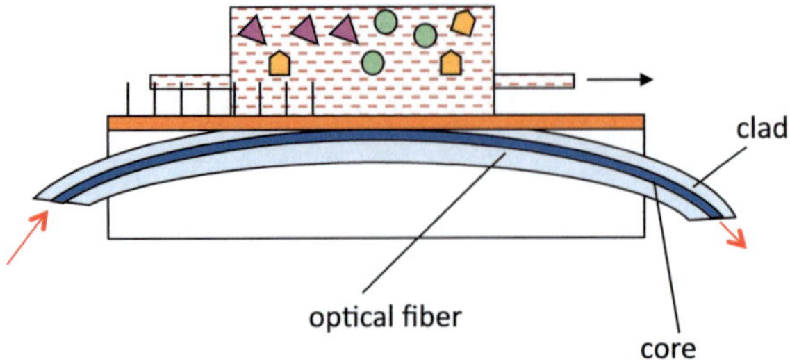

Fig. 3.14 Optical fiber propagating surface plasmon resonance biosensor [28]

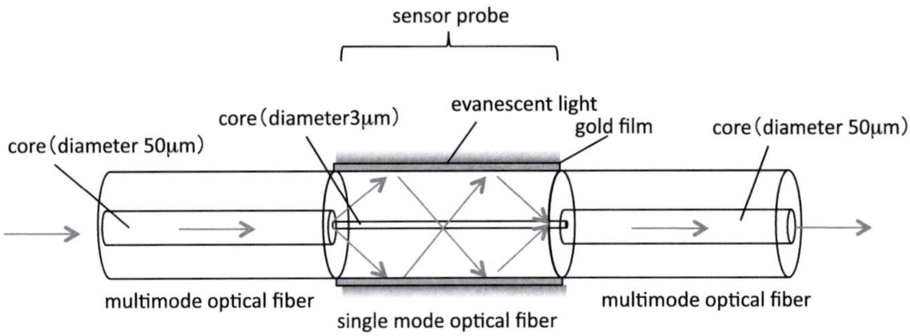

Fig. 3.15 Hybrid type optical fiber surface plasmon resonance biosensor [29]

on the gold thin film surface for the use of biosensor. When analyte-binding events occur, the resonance frequency is shifted. From the amount of the shift, one can deduce the amount of the binding analyte. The merit of this method is free from mechanics and easy to handle. However, the optical fiber is not mechanically strong and gentle operation is necessary. To prevent this, a hybrid optical fiber of single and multimode fibers is proposed as shown in Fig. 3.15 [29]. Light is incident to the multimode mode fiber. It is connected to single mode fiber, and then the light propagates inside the clad of single mode optical fiber because of mismatching of the core size. Since diameter of the core of the single-mode fiber is rather large, it is mechanically strong. Another merit of this sensor is that it is not necessary to remove the clad of the optical fiber.

Since the surface plasmon resonance biosensor probes refractive index change of the surrounding medium, the smallest change of refractive index that can be probed is a good parameter to evaluate the performance of the sensor. In the case of the ATR-based surface plasmon biosensor, the refractive index resolution is usually in the order of 10^{-5}. The corresponding mass resolution, which is the smallest change of mass that can be probed, is a few $\mathrm{pg/mm^2}$ to $10\,\mathrm{pg/mm^2}$. There are a few methods to improve the performance of the surface plasmon resonance biosensor. Among them, measurements of phase of reflected light are promising. There are two method to measure the phase of the reflection light. One is the heterodyne detection and the other is surface plasmon ellipsometry. In the heterodyne detection, two Zeeman lasers are

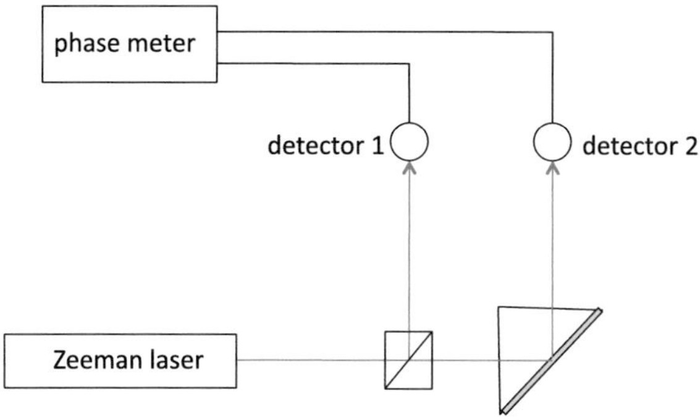

Fig. 3.16 Hybrid type optical fiber surface plasmon resonance biosensor

used [30–33]. The lasers emit light at slightly different frequency Δf with orthogonal polarizations at intermediate polarization between p- and s-polarizations. The optical geometry is shown in Fig. 3.16. The intensities of reflected light at detector 1, I_1, and at detector 2, I_2, with respect to time t are expressed by

$$I_1 = \cos(\Delta f t + \phi_0) \tag{3.21}$$

and

$$I_2 = \cos(\Delta f t + \Delta\phi + \phi_0), \tag{3.22}$$

where ϕ_0 is the initial phase and ϕ is the difference of the phase. By use of the phase detection apparatus, we have the phase difference $\Delta\phi$, which is affected by the refractive index of ambient medium. The report shows high phase sensitivity of the order of 10^{-5} degrees, and the refractive index resolution of 10^{-7}.

Another method, surface plasmon ellipsometry, also provides good index resolution [34, 35]. The method can be made by installing a polarizer between the sample and the detector, used as an analyzer, as shown in Fig. 3.17. The polarization of incident light is rotated by a certain degrees with respect to the p-polarization. When the analyzer is rotated, the profile of reflected light is sinusoidal with an offset with respect to the analyzer angle A, as shown in Fig. 3.18. Fitting the profile

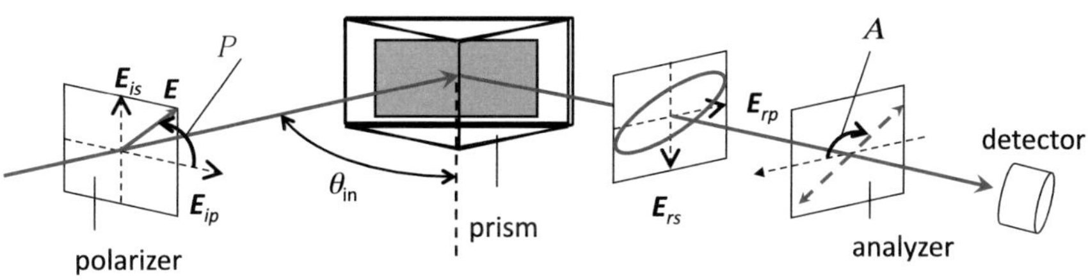

Fig. 3.17 Phase measurement by surface plasmon ellipsometry

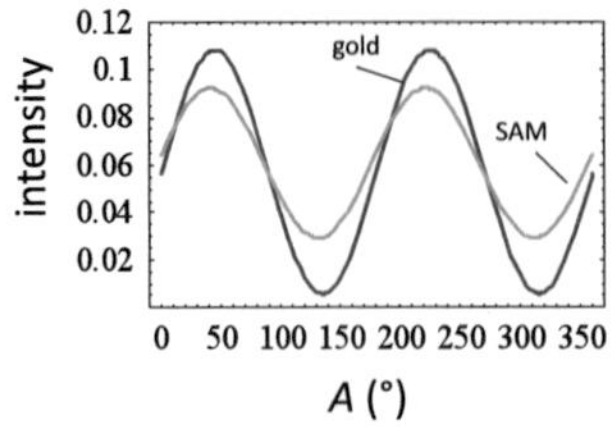

Fig. 3.18 Typical intensity profile in surface plasmon ellipsometry

to the theoretical curve gives the phase difference $\Delta\phi$, according to the following expression:

$$I(A) \sim 1 + \sqrt{\alpha^2 + \beta^2}\sin(2A + \phi), \tag{3.23}$$

where α and β can be determined from the reflectivity profile. The index resolution σ_{RI} is calculated with the following equation [36]

$$\sigma_{\mathrm{RI}} = \sigma_{\mathrm{i}}\frac{1}{\eta}, \tag{3.24}$$

where η is sensitivity, and σ_{i} is the instrument resolution. In the phase measurement, it is

$$\eta = \frac{d\phi}{dn}. \tag{3.25}$$

Table 3.1 summarizes the parameters for different surface plasmon biosensors [35]. Parameters for the resonance angle measurement is also given fore reference. The refractive index resolution of both phase measurements are in the order of 10^{-7}, which is a considerably good value.

3.4.3 LSP Biosensors

We can use localized surface plasmon resonance for biosensors. Since localized surface plasmons occurs in a metallic particle, we can fabricate a biosensors in the order of nanometers, in principle. In addition, we do not have to use rotational stage,

Table 3.1 Refractive index resolution for different surface plasmon biosensors

technique (mm)	η (RI/deg)	σ_{i} (deg)	σ_{RI}
Resonance angle	9.1×10^{-3}	0.001	9.1×10^{-6}
Heterodyne	2.2×10^{-5}	0.01	2.2×10^{-7}
Rotation analyzer ellipsometry	2.2×10^{-5}	0.02	4.4×10^{-7}

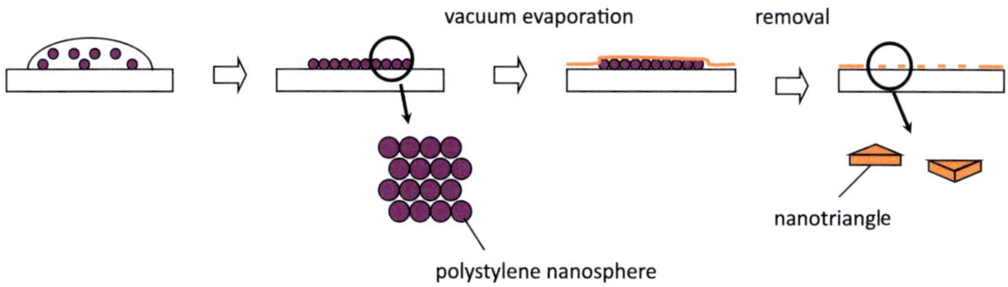

Fig. 3.19 The nanosphere lithography

so that compact biosensors can be made. In many researches, gold nano spheres are used. We can also use core-shell particles, rough surfaces and gold island particles. Also lithography technique can be used for the formation of the nano particles.

Localized surface plasmon biosensors are firstly proposed in 2000. Okamoto et al. proposed a localized surface plasmon sensor using gold nanospheres immobilized on a glass substrate [37]. Poly(methylmethacryalte) (PMMA) layers with different thicknesses were deposited on the substrate by the spin-coating method. They measured the thicknesses of PMMA layers on the substrate by absorption spectra. The absorption peaks of gold nanospheres are shifted from 512 to 540 nm with increasing the thickness of the PMMA layer. Although the experiments are not for probing biological molecules, the results clearly suggest the application of this method to biosensors. In 2002, a few localized surface plasmon biosensors are reported. Nath and Chilkoti reported detection of avidin, by use of the gold nanospheres immobilized on a glass substrate [38–40]. Biotin is immobilized on the gold nanosphere surface, which has affinity to avidin. They measured transmission spectra and observed that the peak wavelength is red-shifted after the avidin binding. The limit of detection of avidin was 0.83 nM. They improved this type of biosensor using gold nanorods and achieved a good detection limit of 94 pM [41]. Haes and Van Duyne reported localized surface plasmon resonance biosensor using the triangular silver nanoparticles using nanosphere lithography [42, 43]. The scheme of the nanosphere lithography is illustrated in Fig. 3.19. First, a polystyrene nanosphere monolayer with a diameter of about 400 nm are deposited on a substrate. After deposition of silver on the substrate by vacuum evaporation, the polystyrene nanospheres are removed. Then nanotriangles are left on the surface which behave as a localized surface plasmon resonance biosensor. They achieved a quite good detection limit of avidin in the order of 10^{-12} M. A localized surface plasmon biosensor of the other type is reported by Takei [44]. They deposited polystyrene nanospheres on a gold substrate and they deposit a gold thin by vacuum evaporation. The core (the polystyrene nanosphere)-shell (the thin gold film) nanoparticles work as a sensitive biosensor.

When gold or silver is deposited on a substrate at a thickness of a few nanometers, the film is not continuous and is composed of islands. By annealing process, the metallic islands becomes hemispheres or truncated spherical shape. These nanoparticles can be also used as a biosensors. Meriaudeau et al. [45] and Rubinstein et al. [46] used these island films and performed biosensing. The advantage of this method

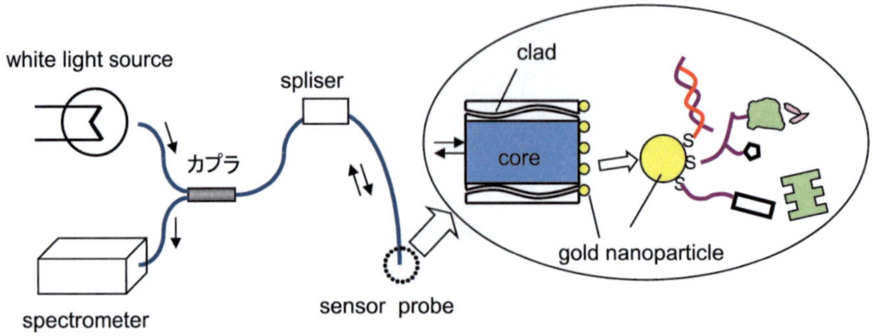

Fig. 3.20 Localized surface plasmon optical fiber biosensor

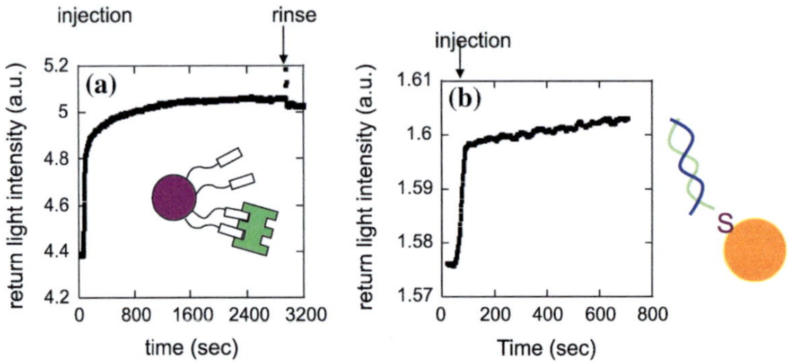

Fig. 3.21 **a** Protein (avidin) and **b** DNA detection

is that it is easy to form the gold nanoparticles. The disadvantage is that it is difficult to form particles with homogenous sizes and shapes. Nanoholes in a continuous metallic film also work as a biosensor. Höök et al. reported localized surface plasmon biosensor of nanoholes with 140 nm diameter and 15–25 nm in depth [47]. The index sensitivity of 270 nm/RIU is reported. Other group has reported the mass resolution of 1 pg/mm^2 [48].

A type of biosensor that utilizes the merit of the localized surface plasmons is an optical fiber biosensor. Mitsui et al. reported an optical fiber biosensor in which gold nano particles are deposited at an endface of the optical fiber, as shown in Fig. 3.20 [49]. The merits of this sensor are (a) high mechanical strength, (b) low cost, (c) small sample volume, and (d) easy to handle. Figure 3.21a, b shows the reflective light intensity when a protein (avidin) or DNA sample solution was injected, as a function of incubation time. The increase in the reflection light intensity corresponds to binding of analyte to ligand. The achieved a refractive index sensitivity was 2×10^{-5}, and the detection of proteins and DNAs with a sample volume of 50 nL.

3.4.4 Surface Plasmon Microscope

The surface plasmon microscope images the difference in the reflectivity due to existence of molecules. It is firstly reported by Rothenhäusler and Knoll [50]. The optical setup is depicted in Fig. 3.22. The setup is similar to the usual propagating surface plasmon biosensors based on ATR, except for using CCD camera and a lens. The area where monolayers or molecules exist has the surface plasmon resonance condition which is different from the bare gold substrate. Hence the reflectivity is different when the angle of incidence is fixed at the near resonance angle. Figure 3.23 shows the images. In Fig. 3.23a, the ATR scan profile with the solid line shows the reflectivity at the gold surface without any monolayer and that with broken line shows the reflectivity at the area of a monolayer. At the angle of incidence at θ_1, the monolayer area has higher reflectivity, at θ_2 the reflectivity is the same in both area and the monolayer area has lower reflectivity at θ_3. The corresponding images are shown in Fig. 3.23a–c. This microscope is successfully applied to a multichannel biosensor [51–53], with which the DNA detection was demonstrated.

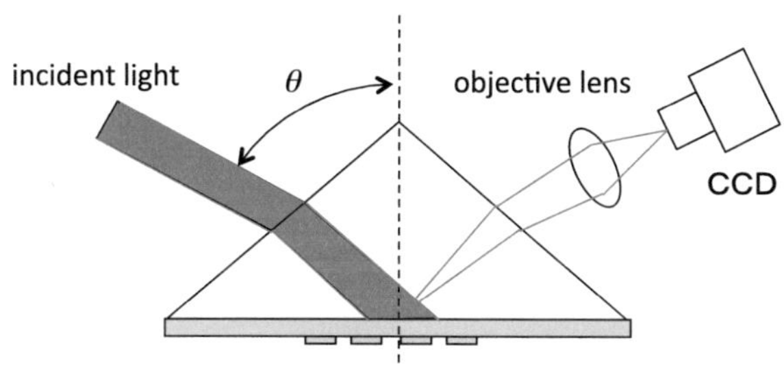

Fig. 3.22 Optical setup of surface plasmon resonance microscope

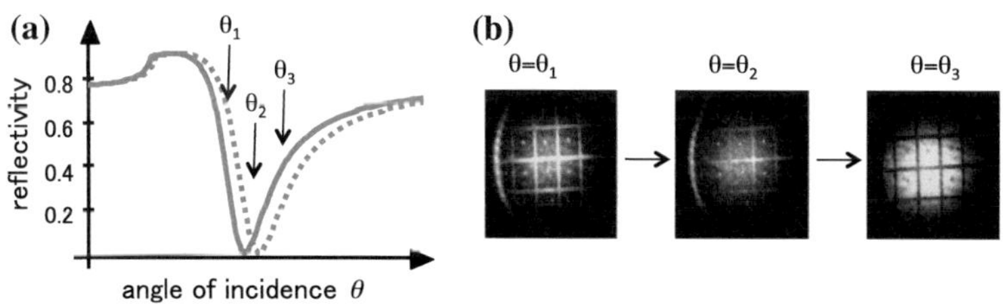

Fig. 3.23 **a** ATR profile and **b** the corresponding surface plasmon resonance microscope images

3.5 Surface Enhanced Raman Scattering Spectroscopy

3.5.1 Introduction

Raman scattering spectroscopy is a kind of vibration spectroscopy. It is a phenomenon that the scattered light is frequency-shifted by the vibration frequency $\Delta\nu$ of a group in a molecule. This is schematically shown in Fig. 3.24. Light at frequency ν is incident to molecules, the strongest component of scattered light is Rayleigh scattering at the same frequency. In addition, two components of Raman scattering can be observed. The scattered light at a lower frequency $\nu - \Delta\nu$ is called Stokes light, and that at a higher frequency $\nu + \Delta\nu$ called anti-Stokes light. Since the vibration frequency $\Delta\nu$ is specific for groups in a molecule, this method gives information of species or state of molecules. A remarkable feature of the Raman scattering spectroscopy is that excitation and emission are in the visible or near-infrared range, although the vibration frequency is in the mid-infrared range. A laser light source for excitation and a low noise detector can be used. Thus the signal-to-noise ratio is much better than infrared spectroscopy. On the other hand, it has a very small scattering cross-section in the order of 10^{-30} cm^2 [54]. It is the demerit of Raman scattering spectroscopy that approximately 10^8 molecules are necessary for detection. Even in resonant Raman scattering spectroscopy for probing dye molecules, the cross-section is better in the order of 10^{-26} cm^2, and is still much less than fluorescence spectroscopy, whose cross-section is 10^{-18} cm^2.

In 1977, Albrecht and Creighton [55], and Van Duyne and Jeanmaire [56] independently reported the huge enhancement of Raman signal from molecules on a metal. These reports are the beginning of Surface enhanced Raman scattering spectroscopy (SERS) spectroscopy, which provides an extremely strong Raman scattering signal. It is, therefore, a powerful tool to probe a small amount of molecules adsorbed on metallic nanostructures.

Here we show a SERS spectrum. Figure 3.25 shows Raman spectra of Rhodamine 6G adsorbed on (a) smooth gold surface, (b) gold nanospheres immobilized on a

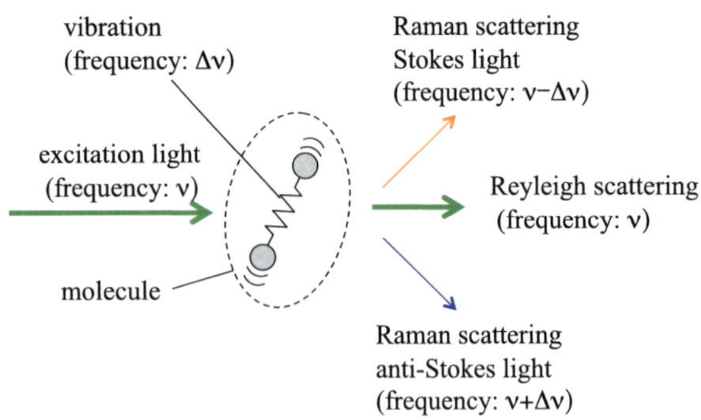

Fig. 3.24 Schematic of Raman scattering

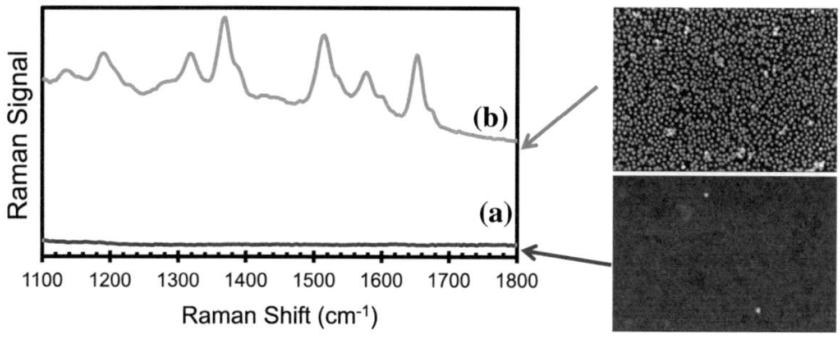

Fig. 3.25 Raman spectra of Rhodamine 6G adsorbed on (**a**) a smooth gold surface (**b**) gold nanospheres. The spectra are provided by Mr. Yusuke Nagai of Tokyo institute of technology

gold surface. The scanning electron microscope images of the gold surfaces are also shown. While the Raman spectrum is absent in (a), it is observed in (b). The large enhancement of localized surface plasmons allows us to obtain the spectrum.

The extremely large enhancement of SERS spectroscopy is ascribed to two reasons: the physical and chemical. The physical reason is attributed to the large electric field produced on a metallic nanostructures, which originates from localized surface plasmons. The chemical reason is attributed to the rise of vibration due to charge transfer or electrical coupling between the molecules to the metal. In general, the former reason is much larger as described above, while the latter gives an enhancement of 10–10^2 [57, 58]. Hence, the researches on SERS enhancement concentrate on the creation of metallic nanostructures effective for SERS.

In early works, silver colloid aggregates or rough surfaces were used for SERS. Silver is not a stable metal in atmosphere. It is easily oxidized or sulfated, and loses its metallic character. In recent studies, gold surfaces or nanostructures are sometimes used because of the better chemical stability. In addition, electron beam lithography or the focused ion beam process can be used to make the gold nanostructures in a desired shape. Therefore many works have appeared on the SERS spectroscopy in the last decade [59].

3.5.2 *Raman Intensity*

As is described above, the physical reason is important to understand the SERS enhancement. We will consider the SERS enhancement from the electromagnetic point of view. When light with an electric field, $\overrightarrow{E}_0$, is incident to an object with a Raman tensor $\overleftrightarrow{\chi}$, the Raman scattering light intensity I_R is approximated to

$$I_\mathrm{R} = C((\overleftrightarrow{L}_\mathrm{R}\hat{e}_\mathrm{out}) \cdot \overleftrightarrow{\chi}\,\overleftrightarrow{L}_0\overrightarrow{E}_0)^2, \tag{3.26}$$

Fig. 3.26 The reciprocal theorem

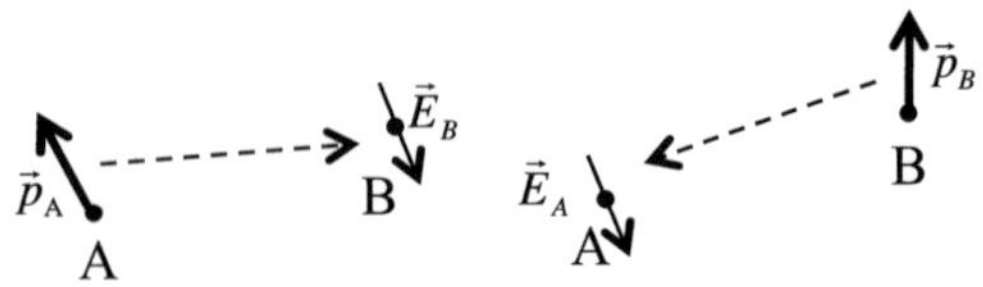

where $\overleftrightarrow{L}_0$ and $\overleftrightarrow{L}_R$ are local field factor tensors for incident and Raman frequency, $\hat{e}_{out}$ is the unit vector of output polarization, and C is a constant. The local field factor tensor gives a local electric field, $\vec{E}_{loc}$ produced by an incident light field, $\vec{E}_0$, as

$$\vec{E}_{loc} = \overleftrightarrow{L}_0 \vec{E}_0. \tag{3.27}$$

If an enhanced electric field is absent, $\overleftrightarrow{L}$ can be regarded as the identity matrix. However, SERS is not the case, because of the LSPs at metallic surfaces or nanostructures.

Equation 3.26 is derived from the reciprocal theorem for the electric field $\vec{E}$ produced by a dipole $\vec{p}$ [60]. Suppose that a dipole $\vec{p}_A$ located at A produces an electric field at B, $\vec{E}_B$, as shown in Fig. 3.26. According to the reciprocal theorem, the relation is described as $\vec{p}_A \cdot \vec{E}_A = \vec{p}_B \cdot \vec{E}_B$. This relation holds in the framework of dipole radiation.

According to Eq. 3.26, the signal is almost proportional to the fourth power of the local electric field. Hence an enhancement of the local electric field due to localized surface plasmons is extremely effective. When the electric field of excitation light is enhanced ten times, the Raman intensity is 10^4 times higher. At a narrow gap of two metallic colloid dimer, nearly 100-fold enhancement of an electric field of light is obtained. Then the estimated enhancement is 10^8. In addition to resonance Raman effect, we have a SERS cross-section in the order of 10^{-16}–10^{-18} cm^2 for dye molecules at optimized conditions. This cross-section is similar to the fluorescence spectroscopy. Single-molecule SERS spectroscopy is reported in 1997 from the dimer system [11, 61, 62], and a number of reports on single-molecule SERS have been reported after that.

3.5.3 Calculation of SERS Intensity

For designing metallic nanostructure with high SERS efficiency, the calculation methods of SERS intensity are necessary. There are a few calculation methods to evaluate the electric fields of light around a metallic nanostructure. The finite-difference time-domain (FDTD) method is the most common tool [63]. It is a kind of difference method for time and space. The algorithm is relatively simple and a nanostructure in any shape can be calculated. There are many commercially available softwares. The finite element method can be used for the calculation. Discrete dipole approximation

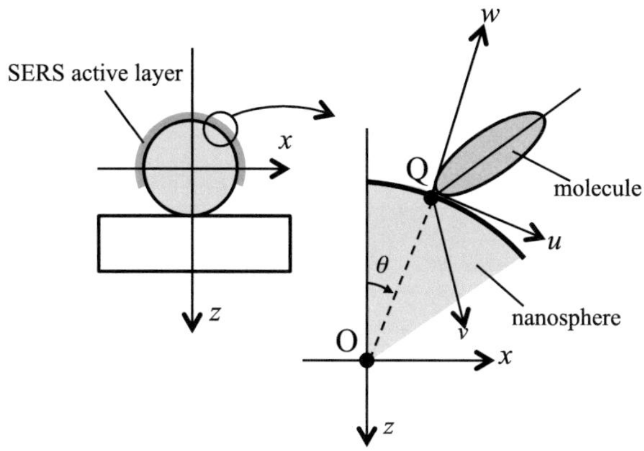

Fig. 3.27 Geometry for considering excitation and emission from a dipole (molecule)

(DDA) is another type of a calculation method [64, 65]. Nanostructures are divided into an array of spheres with a polarizability α, and the optical response is calculated including the interaction between them, considering the retardation effects. The interaction is considered through the induced dipole p_{ind}, which is produced as $p_{\mathrm{ind}} = \alpha E_{\mathrm{loc}}$ by the local electric field E_{loc}. This method also can be applied to a nanostructure in any shape. This method is originally developed for calculating interstellar dust grains [64]. There are a couple of open softwares available.

The FDTD and DDA methods provide the local electric field $\overrightarrow{E}_{\mathrm{loc}}$. By comparing $\overrightarrow{E}_{\mathrm{loc}}$ to the incident light field $\overrightarrow{E}_0$, we have local field factor tensor $\overleftrightarrow{L}$. However the reciprocal theorem is not always valid. In Eq. 3.26, while the excitation part, $\overrightarrow{E}_{\mathrm{loc}} = \overleftrightarrow{L}_0 \overrightarrow{E}_0$, always holds, the emission part, $(\overleftrightarrow{L}_{\mathrm{R}} \hat{e}_{\mathrm{out}}) \overrightarrow{p}$, does not always hold, where $\overrightarrow{p}$ is the emission source diple. Hence we need to verify the validity of the reciprocal theorem for each object [60].

Let us consider a simple case that a dipole (molecule) $\overrightarrow{p}_{\mathrm{Q}}$ is located at point Q on a gold nanosphere with a radius R that is much smaller than the wavelength of light, as shown in Fig. 3.27a. Hence, for simplicity, we use the quasi-static approximation, which is also called the long-wavelength approximation. The local field $\overrightarrow{E}_{\mathrm{loc}}$ at Q is described with a local field tensor $\overleftrightarrow{L}_0$ and incident field $\overrightarrow{E}_0$, as described in Eq. 3.27. The local field is composed of two contributions, one the incident electric field, $\overrightarrow{E}_0$, and the other an electric field from the induced dipole at the center of the nanosphere [66]. The latter is described as $\overleftrightarrow{X}\,\overleftrightarrow{\alpha}\,\overrightarrow{E}_0$, where $\overleftrightarrow{X}$ is an operator that gives the electric field at the surface, produced by a dipole at the center of the nanosphere. It is written by

$$\overleftrightarrow{X} = \frac{1}{R^3}\begin{pmatrix} -1 & 0 & 0 \\ 0 & -1 & 0 \\ 0 & 0 & 2 \end{pmatrix}. \tag{3.28}$$

Then the local electric field at Q is described as

$$\overrightarrow{E}_{\text{loc}} = (\overleftrightarrow{X}\,\overleftrightarrow{T}\,\overleftrightarrow{\alpha} + \overleftrightarrow{T})\overrightarrow{E}_0, \tag{3.29}$$

where $\overleftrightarrow{T}$ is the transform matrix from the laboratory coordinates (x, y, z) to the local coordinates (u, v, w) at point Q, as shown in Fig. 3.27b. Here the local field tensor for excitation is

$$\overleftrightarrow{L}_0 = \overleftrightarrow{X}\,\overleftrightarrow{T}\,\overleftrightarrow{\alpha} + \overleftrightarrow{T}. \tag{3.30}$$

The emission part can be described similarly: the direct emission from the source dipole $\overrightarrow{p}$ and emission from the nanosphere. Then the electric field the detector E_{R} is

$$\overrightarrow{E}_{\text{R}} = (\overleftrightarrow{\alpha}\,\overleftrightarrow{T}^{-1}\overleftrightarrow{X} + \overleftrightarrow{T}^{-1})\overrightarrow{p}. \tag{3.31}$$

The local field tensor for emission is

$$\overleftrightarrow{L}_{\text{R}} = \overleftrightarrow{\alpha}\,\overleftrightarrow{T}^{-1}\overleftrightarrow{X} + \overleftrightarrow{T}^{-1}. \tag{3.32}$$

Since $\overleftrightarrow{T}$ is not a diagonal tensor, $\overleftrightarrow{L}_{\text{R}}$ is not necessarily equal to $\overleftrightarrow{L}_0$. This means that the reciprocal theorem does not hold. Therefore we calculate the SERS intensity from an infinitesimal element $\sin\theta d\theta d\phi$ at a point Q, using the following equation.

$$I_R \sin\theta d\theta d\phi = C\langle\left|\hat{e}_{\text{out}} \cdot \overleftrightarrow{L}_{\text{R}} \cdot \overleftrightarrow{\chi} \cdot \overrightarrow{E}_{\text{loc}}(r, \theta, \phi)\right|^2_{r=R}\rangle d\phi \sin\theta d\theta. \tag{3.33}$$

The bracket denotes the average over the distribution of the molecules. The total intensity I_{SERS} is

$$I_{\text{SERS}} = \int_S I_R dS, \tag{3.34}$$

where S is the area covered by the SERS active molecule, and $dS = \sin\theta d\theta d\phi$. By use of these expression, a simulation of the angular-resolved SERS spectroscopy has been carried out, and compared the experimental results, in order to probe the spatial molecular distribution on a gold nanosphere [66].

References

1. H. Raether, *Surface Plasmons on Smooth and Rough Surfaces and on Gratings* (Springer, Heidelberg, 1988)
2. W. Knoll, Makromol. Chem. **192**, 2827 (1991)
3. W. Knoll, MRS Bull. **16**, 29 (1991)
4. B. Lamprecht, J.R. Krenn, G. Schider, H. Ditlbacher, M. Salerno, N. Felidj, A. Leitner, F.R. Aussenegg, Appl. Phys. Lett. **79**, 51 (2001)
5. E. Knestschman, H. Raether, Z. Naturforsch. **23a**, 2135 (1968)
6. A. Otto, Z. Phys. **216**, 398 (1968)

7. B. Liedgurg, C. Nylander, I. Pundstrom, Sens. Actuators **4**, 299 (1983)
8. J. Homola, S.S. Yee, G. Gauglitz, Surface plasmon resonance sensors: review. Sens. Actuators B **54**, 3 (1999)
9. J. Homola, Chem. Rev. **108**, 462 (2008)
10. W. Knoll, Annu. Rev. Phys. Chem. **49**, 569 (1998)
11. K. Kneipp, H. Kneipp, I. Itzkan, R.R. Dasari, M.S. Feld, Chem. Rev. **99**, 2957 (1999)
12. N. Hayazawa, Y. Inouye, Z. Sekkat, S. Kawata, Opt. Commun. **183**, 333 (2000)
13. J. Takahara, S. Yamagishi, H. Taki, A. Morimoto, T. Kobayashi, Opt. Lett. **22**, 475 (1997)
14. C.F. Bohren, D.R. Huffman, *Absorption and Scattering of Light by Small Particles* (Wiley, New York, 1983)
15. P.B. Johnson, R.W. Christy, Phys. Rev. B **6**, 4370 (1972)
16. H. Wang, T.B. Huff, D.A. Zweifel, W. He, P.S. Low, A. Wei, J.-X. Chen, Proc. Nat. Acad. Sci. U S A **102**, 15751 (2005)
17. N.J. Durr, T. Larson, D.K. Smith, B.A. Korgel, K. Sokolov, A. Ben-Yakar, Nano Lett. **7**, 941 (2007)
18. U. Mahamood, R. Weissleder, Mol. Cancer Ther. **2**, 489 (2003)
19. S. Lal, S.E. Clare, N.J. Halas, Acc. Chem. Res. **41**, 1842 (2008)
20. A.E. Neeves, M.H. Birnboim, J. Opt. Soc. Am. B **6**, 787 (1989)
21. C. Loo, A. Lowery, N. Halas, J. West, R. Drezek, Nano Lett. **5**, 709 (2005)
22. A.M. Gobin, M.H. Lee, N.J. Halas, W.D. James, R.A. Drezek, J.L. West, Nano Lett. **7**, 1929 (2007)
23. P.K. Aravind, H. Metiu, Surf. Sci. **124**, 506 (1983)
24. R. Ruppin, J. Phys. Soc. Jpn. **58**, 1446 (1989)
25. M.M. Wind, J. Vlieger, D. Bedeaux, Phys. A **141**, 33 (1987)
26. F. Garuso, K. Niikura, D.N. Furlong, Y. Okahata, Langmuir **13**, 3427 (1997)
27. Y. Manaka, Y. Kudo, H. Yoshimine, T. Kawasaki, K. Kajikawa, Y. Okahata, Chem. Commun. 3574 (2007)
28. R. Slavik, J. Homola, J. Ctyroky, E. Brynda, Sens. Actuators B **74**, 106 (2001)
29. M. Iga, A. Sekib, K. Watanabe, Sens. Actuators B **106**, 363 (2005)
30. K. Chen, C. Hsu, D.S. Chen, C. Hsu, D. Su, Opt. Eng. **42**, 1884 (2003)
31. J. Guo, Z. Zhu, W. Deng, S. Shen, Opt. Eng. **37**, 2998 (1998)
32. C. Wu, Z. Jian, S. Joe, L. Chang, Sens. Actuators B **92**, 133 (2003)
33. Y. Xinglong, Z. Lequn, J. Hong, W. Haojuan, Y. Chunyong, Z. Senggeng, Sens. Actuators B **76**, 199 (2001)
34. F. Abele, T. Lopez-Rios, A. Tadjeddine, Solid State Commun. **16**, 843 (1975)
35. R. Naraoka, K. Kajikawa, Sens. Actuators B (Chemical) **107**, 952 (2005)
36. S.G. Nelson, K.S. Johnston, S.S. Yee, Sens. Actuators B **35–36**, 187 (1996)
37. T. Okamoto, I. Yamaguchi, T. Kobayashi, Opt. Lett. **25**, 372 (2000)
38. N. Nath, A. Chilkoti, Anal. Chem. **74**, 504 (2002)
39. N. Nath, A. Chilkoti, J. Fluoresc. **14**, 377 (2004)
40. N. Nath, A. Chilkoti, Anal. Chem. **76**, 5370 (2004)
41. S.M. Marinakos, S. Chen, A. Chilkoti, Anal. Chem. **79**, 5278 (2007)
42. A.J. Haes, R.P. Van Duyne, J. Am. Chem. Soc. **124**, 10596 (2002)
43. J. Zhao, X. Zhang, C.R. Yonson, A.J. Haes, R.P. Van Duyne, Nanomedicine **1**, 219 (2006)
44. M. Himmelhaus, H. Takei, Sens. Actuators B **63**, 24 (2000)
45. F. Meriaudeau, T. Downey, A. Wig, A. Passian, M. Buncick, T.L. Ferrell, Sens. Actuators B **54**, 106 (1999)
46. G. Kalyuzhny, A. Vaskevich, M.A. Schneeweiss, I. Rubinstein, Chem. A Eur. J. **8**, 3849 (2002)
47. A.B. Dahlin, J.O. Tegenfeldt, F. Höök, Anal. Chem. **78**, 4416 (2006)
48. D. Gao, W. Chen, A. Mulchandani, J.S. Schultz, Appl. Phys. Lett. **90**, 073901 (2007)
49. K. Mitsui, Y. Handa, K. Kajikawa, Appl. Phys. Lett. **85**, 4231 (2004)
50. B. Rothenhausler, W. Knoll, Nature **332**, 615 (1988)
51. J.M. Brockman, B.P. Nelson, R.M. Corn, Annu. Rev. Phys. Chem. **51**, 41 (2000)
52. J.M. Brockman, A.G. Frutos, R.M. Corn, J. Am. Chem. Soc. **121**, 8044 (1999)

53. I. Mannelli, L. Lecerf, M. Guerrouache, M. Goossens, M.-C. Millot, M. Canva, Biochem. Bioelectron. **22**, 803 (2007)
54. B. Pettinger, K. Krischer, J. Electron. Spectrosc. Relat. Phenom. **45**, 133 (1987)
55. M.G. Albrecht, J.A. Creighton, J. Am. Chem. Soc. **99**, 5215 (1977)
56. D.L. Jeanmaire, R.P. Van Duyne, J. Electroanal. Chem. **84**, 1 (1977)
57. A. Otto, J. Raman Spectrosc. **36**, 497 (2005)
58. W.-H. Park, Z.H. Kim, Nano Lett. **10**, 4040 (2010)
59. E.C. Le Ru, P.G. Etchegoin, *Principles of Surface-Enhanced Raman Spectroscopy: and related plasmonic effects* (Elsevier, Amsterdam, 2009)
60. E.C. Le Ru, P.G. Etchegoin, Chem. Phys. Lett. **423**, 63 (2006)
61. S. Nie, S.R. Emory, Science **275**, 1102 (1997)
62. K. Kneipp, Y. Wang, H. Kneipp, L.T. Perelman, I. Itzkan, R.R. Dasari, M.S. Feld, Phys. Rev. Lett. **78**, 1667 (1997)
63. K.S. Yee, IEEE Trans. Antennas Propag. **14**, 302 (1966)
64. E.M. Purcel, C.R. Pennyacker, Astrophys. J. **186**, 705 (1973)
65. B.T. Draine, P.J. Flatau, J. Opt. Soc. Am. A **11**, 1491 (1994)
66. Y. Nagai, T. Yamaguchi, K. Kajikawa: J. Phys. Chem. C **116**, 9716 (2012)

Chapter 4
Optical Metamaterials

Kotaro Kajikawa

Abstract Optical metamaterials show interesting optical properties originating from their higher-order structures. We can obtain optical media with a wide range of refractive indexes: from negative to positive large values. This allows us to realize special optical functions such as super high resolution microscopy and optical croaking. In future, various useful optical properties will be found or created in optical metamaterials. This chapter deals with the concept of the optical metamaterials and some optical applications.

4.1 Introduction

Optical metamaterials [1] show high functionality originating from their higher order structures. Similar things are found in proteins in a biological system that the functionality is due to their higher-order structure. Metamaterials start to be investigated at microwave frequencies. The original concept was that an assembly of electrical circuits is compared to a matter, since the electromagnetic response from the electrical circuits, such as LC resonators, can mimic the response from the matter. In the last decade, this concept is extended to light frequencies. Negative index materials and the optical cloaking technique at optical frequencies are realized, not only because of the development of micro and nano fabrication techniques but also the development of design for metamaterials. The basic concept to realize the metamaterials is the control of permittivity and permeability by designing the nanostructures. In addition, surface plasmons play an important role to control its permittivity. It is sure that metamaterials will be an important research field as photonic materials in future.

K. Kajikawa (✉)
Interdisciplinary Graduate School of Science and Engineering, Tokyo Institute of Technology, Nagatsuta, Midori-ku, Yokohama, 226-8502 Japan
e-mail: kajikawa@ep.titech.ac.jp

Y. Aoyagi and K. Kajikawa (eds.), *Optical Properties of Advanced Materials*,
Springer Series in Materials Science 168, DOI: 10.1007/978-3-642-33527-3_4,

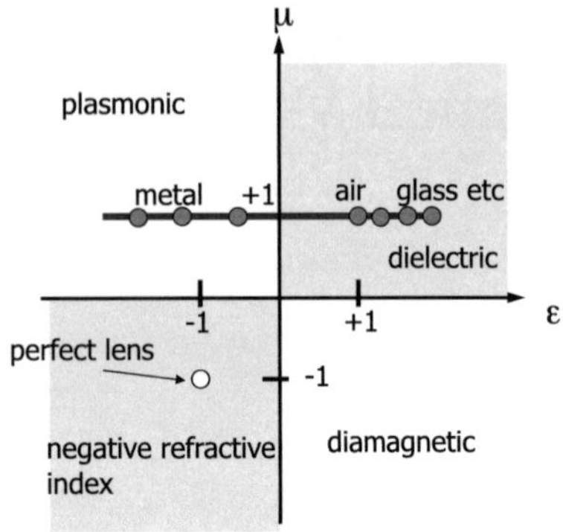

Fig. 4.1 Plots of existing matters and metamaterials in $\epsilon - \mu$ plain

4.2 Metamaterials and Meta-Molecules

Metallic nanostructures much smaller than the optical wavelength sometimes show interesting optical properties that are not found in existing bulk materials. Most of them are ascribed to surface plasmons that sometimes cause a huge polarizability at their resonance wavelength. Metallic nano-materials allow us to control not only permittivity but also permeability because of the fast response of free electrons in a metal. One of them is negative refractive index. The refractive index n of a matter is written by

$$n = \pm\sqrt{\epsilon\mu} \tag{4.1}$$

where ϵ and μ are its permittivity and permeability, respectively. The permeability of an existing material is unity because the magnetic response of a matter is absent at optical frequencies. This is because the magnetism comes from the orbital or spin angular momentum, whose response time is below the frequency range of microwaves. On the other hand, the permittivity of existing materials is ranging from -25 to 16. As a result, the real parts of the refractive indexes of the existing materials are ranging from 0 to 4 at optical frequency. This is shown in Fig. 4.1, in which existing materials are plotted. They are plotted on the line of $\mu = 1$.

The sign of the refractive index is determined by the sign of the permittivity and permeability [1–4]. If both are negative, the refractive index is negative. Such materials are called "left handed materials (LHM)" or "negative index materials (NIM)". Metamaterials and some kinds of photonic crystals have possibility to be LHM [5]. If we have LHM at optical frequency, incomprehensible phenomena will appear such as negative refraction and super resolution.

An interesting thing is that a medium with negative permeability or permeability of other than unity can be made of existing materials with permeability of unity. As is described above, the magnetism is produced by rotational movement of electrons, which produces angular momentum. Similar effect will appear by the current flow in a circular coil. We can achieve $\mu \neq 1$ by a current flowing in a resonator shown in

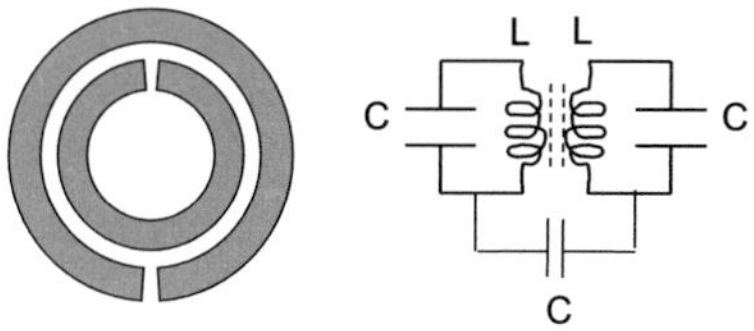

Fig. 4.2 Split ring resonator and its equivalent circuit

Fig. 4.2, whose size is much smaller than the wavelength of light. Such a resonance structure has a permeability μ that is written as

$$\mu(\omega) = 1 - \frac{\omega_{\mathrm{mp}}^2 - \omega_{\mathrm{mo}}^2}{\omega^2 - \omega_{\mathrm{mo}}^2 + i\Gamma_{\mathrm{m}}\omega} \tag{4.2}$$

where ω is the frequency, ω_{mp} is magnetic plasma frequency, ω_{mo} is the magnetic resonance frequency, and Γ_{m} is the damping constant [6]. If the resonance is strong, the permeability μ is ranging from negative to positive. Permittivity also shows similar relation. Therefore we can control the both permittivity and permeability separately by the design of the LC resonance circuit. This is the concept of usual metamaterials. Metamaterials have the refractive index of not only negative values but also a large positive values that the existing materials do not have. Such materials will also show attractive nonlinear optical properties.

Next we consider the similarity between metamaterials and existing matters. The correspondence is summarized in Fig. 4.3. The existing matters can be divided into two categories: one atoms and molecules and the other aggregates and crystals. Atoms and molecules show their own functionality even if interactions between them are absent, namely, they are isolated. This is because their functionality is intrinsic and closed inside the unit. Dye molecules are semiconductor that absorbs light, even if they are isolated in solution. On the other hand, inorganic semiconductors such as silicone show their functionality only in a crystalline from. The isolated atoms do not

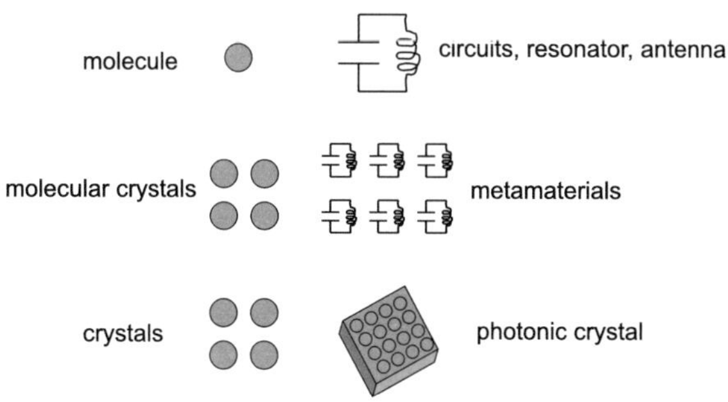

Fig. 4.3 Comparison of existing matters with metamaterials

show the functionality appearing in a crystal. Metamaterials are similar to the atoms and molecules, since it shows optical (electromagnetic) functionality even each unit is separated. Hence, their unit is sometimes called meta-atom or meta-molecules. Metamaterials are an assembly of the meta-atoms or meta-molecules and show the optical properties as an ensemble of the whole units.

Although the metamaterials exhibit their optical functionality of each unit, their aggregates or crystals will also show a new or additional functionality, like molecular aggregates or molecular crystals. An example is that gold bisphere shows a strong red-shifted resonance band, which is not observed in isolated gold nanospheres. As is discussed above, similarity of metamaterials to existing atoms and molecules provides great insight to develop a new class of optical materials.

4.3 Negative Index Materials

One of the interesting optical properties of metamaterials is negative refractive index. Let us consider refraction at a boundary as shown in Fig. 4.4a. Light is incident to medium 2 from medium 1, whose refractive indexes are n_2 and n_1 respectively, at an angle of incidence θ_1. The Snell's law gives the angle of refraction θ_2

$$n_1 \sin\theta_1 = n_2 \sin\theta_2. \tag{4.3}$$

This law is equivalent to conservation of in-plane component of wavevector across the boundary. For negative n_2, the angle of refraction θ_2 is negative, as shown in Fig. 4.4b. This phenomenon is different from diffraction. One may understand the difference when light is traveling in a NIM. Suppose that light is traveling the direction z. In a usual medium, the phase of light is proceeding to the same direction. However, in the case of NIM, the phase of light in the NIM proceeds in the opposite direction. This means that a more distant object looks closer. There are many other interesting phenomena observed in the NIM, exemplified by the inverse Doppler effect and inverse Cherenkov radiation.

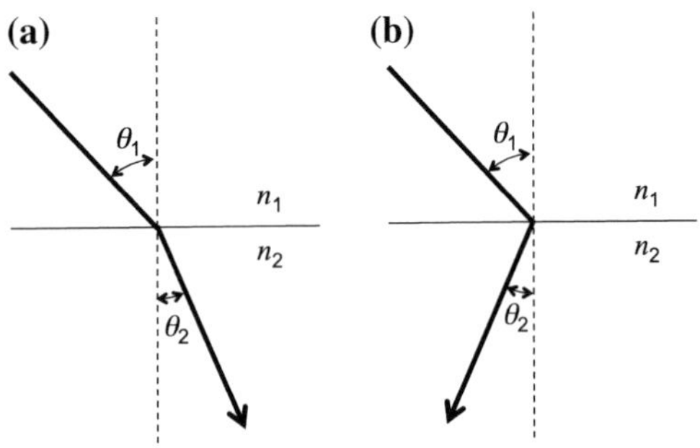

Fig. 4.4 Refraction at boundary. **a** $n > 0$ and **b** $n < 0$

In NIM, both permittivity and permeability are negative. To fabricate such medium, the control of permeability is necessary. Negative permittivity is rather easy, since existing matters such as metals have negative permittivity. In contrast, production of a medium with negative permeability is a thorny problem. At microwave or terahertz frequencies, this can be realized by using an array composed of ring resonators as shown in Fig. 4.3. The dimension of the resonators is in the order of micrometer or millimeters [7, 8]. However it is difficult to fabricate such ring resonators at optical frequencies because of the following two reasons: (a) it is too small and (b) losses in metals are remarkable at optical frequencies. A few structures are proposed instead of the use of the ring resonators. Metallic rod-pairs show negative index in the infrared frequency range [9] and nanometer fishnet structures show negative index at visible wavelengths [10, 11]. The latter structure can be extended to bulky NIM having a thickness. Later, bulky NIMs of a stack of fishnet structure is reported, [12] which work at a wavelength of 1.5 μm and the refraction at negative angles is observed. The optical property appearing as a result of the control the permeability, other than NIM, is a Brewster angle for s-polarized light. This is theoretically predicted [13] and experimentally realized at a microwave frequency [14]

4.4 Effective Medium Approximation

Metamaterials are composite medium composed of metallic nanostructures and matrix media such as air. To calculate the optical constants of metamaterials, such as refractive index and susceptibilities, the homogenization treatment is necessary. The effective medium approximation (EMA) is mostly used. There are four kinds of EMA methods: (a) the parallel capacitor model, (b) serial capacitor model, (c) Maxwell Garnett model and (d) Bruggeman's EMA. Let us consider a composite of two media 1 and 2 with permittivity ϵ_1 and ϵ_2. Suppose the volume fraction f_1 of medium 1. The models of the two-component composite are schematically illustrated in Fig. 4.5. The parallel capacitor model is the simplest approximation. The two media form capacitors, which form a parallel circuit. Then the effective permittivity ϵ_{eff} is

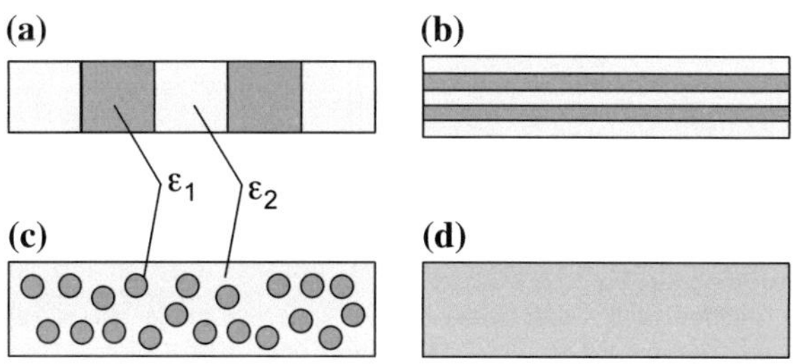

Fig. 4.5 Schematics of (**a**) the parallel capacitor model, (**b**) serial capacitor model, (**c**) Maxwell Garnett model and (**d**) Bruggeman's EMA

described as

$$\epsilon_{\rm eff} = f_1\epsilon_1 + (1 - f_1)\epsilon_2. \tag{4.4}$$

In this model, it is assumed that no interaction exists between the two media. In contrast, the serial capacitor model considers strong interaction between the media. The capacitor from a serial circuit as described in (b). Then the effective permittivity $\epsilon_{\rm eff}$ is described as

$$\epsilon_{\rm eff}^{-1} = f_1\epsilon_1^{-1} + (1 - f_1)\epsilon_2^{-1}. \tag{4.5}$$

In the Maxwell Garnett model, it is assumed that spheres of medium 1 dispersed in medium 2. Then the effective permittivity $\epsilon_{\rm eff}$ is described as

$$\frac{\epsilon_{\rm eff} - 1}{\epsilon_{\rm eff} + 2} = f_1\frac{\epsilon_1 - 1}{\epsilon_1 + 2} + (1 - f_1)\frac{\epsilon_2 - 1}{\epsilon_2 + 2}. \tag{4.6}$$

The Maxwell Garnett model is equivalent to the Claudius-Mosotti relation, used to relate the refractive index and molecular polarizability. The Bruggeman's EMA is proposed as a result of development of the Maxwell Garnett model. In this model the effective permittivity $\epsilon_{\rm eff}$ is described as

$$f_1\frac{\epsilon_1 - \epsilon_{\rm eff}}{\epsilon_1 + 2\epsilon_{\rm eff}} + (1 - f_1)\frac{\epsilon_2 - \epsilon_{\rm eff}}{\epsilon_2 + 2\epsilon_{\rm eff}} = 0. \tag{4.7}$$

Figure 4.6 summarizes the effective permittivity as a function of f_1 using the four different EMAs. In (a), the calculation was made for a dielectric medium with permittivity $\epsilon_1 = 2.5$ is dispersed in a medium 2 with $\epsilon_2 = 1$. In (b) it was made for a metallic medium $\epsilon_1 = -3.17 + 3.11i$ (Au at 510 nm) is dispersed in a medium 2 with $\epsilon_2 = 1$. The calculated results are very different in the case of (b). Although the

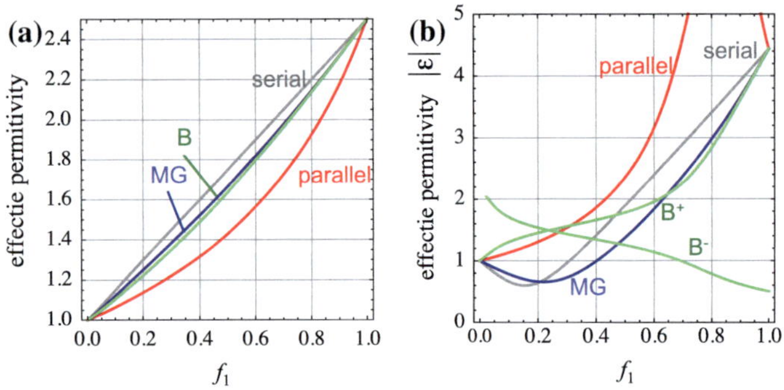

Fig. 4.6 $\epsilon_{\rm eff}$ as a function of f_1 using (i) the parallel capacitor model, (ii) serial capacitor model, (iii) Maxwell Garnett model (MG) and (iv) Bruggeman's EMA (B). **a** A dielectric medium with permittivity $\epsilon_1 = 2.5$ is dispersed in a medium 2 with $\epsilon_2 = 1$. **b** A metallic medium $\epsilon_1 = -3.17 + 3.11i$ (Au at 510 nm) is dispersed in a medium 2 with $\epsilon_2 = 1$

Bruggeman's EMA has two solutions B^+ and B^- in (b), B^+ is real. The application of EMA to metamaterials should be carefully.

4.5 Super Resolution

The super resolution, a resolution over diffraction limit (approximately a half wavelength of light), is one of important applications of NIM. Veselago proposed a perfect lens of NIM, which enables us the super resolution microscopy [2]. The lens is a thin slab of NIM with a refractive index of -1. This utilizes the fact that the angle of refraction is equal to angle of incidence with the opposite sign, as shown in Fig. 4.7. An image of the small object is formed on the other side of the super lesns. In practice, however, this type of super resolution is difficult. The metamaterials cannot be regarded as a continuous medium because the resonance unit of the metamaterials is usually large in the order of 100 nm. Therefore no demonstration of the super resolution was made so far at optical frequencies.

Another super resolution is proposed by Pendry, [3] in which a metallic slab is used, as shown in Fig. 4.8. An object smaller than the wavelength of light forms an image adjacent to the metallic slab through evanescent wave and surface plasmons. Feng et al. demonstrated the super resolution that a chromium thin wire is imaged in a polymer photo-registration film through a 35 nm thick silver film [15]. The structure of the image was probed by atomic force microscope. In this case, the image is not magnified, so that we need to probe the image by the method other than the optical. Zhang et al. reported an optical setup to magnify the image [16]. The geometry is shown in Fig. 4.9. It was demonstrated that two lines separated by 150 nm was imaged using a conventional lens. The observation could be made with a CCD detector. In contrast to the scanning near-field optical microscopy, which also provides optical

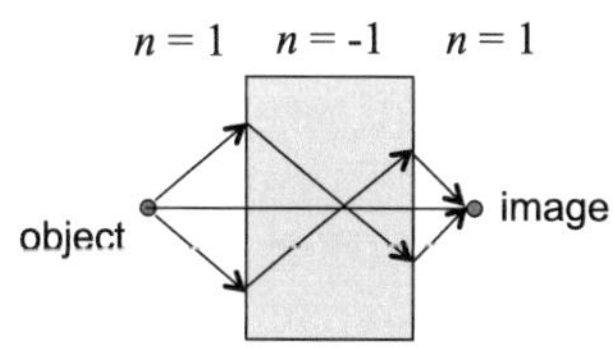

Fig. 4.7 Super resolution with a perfect lens of a NIM slab proposed by Veselago [2]

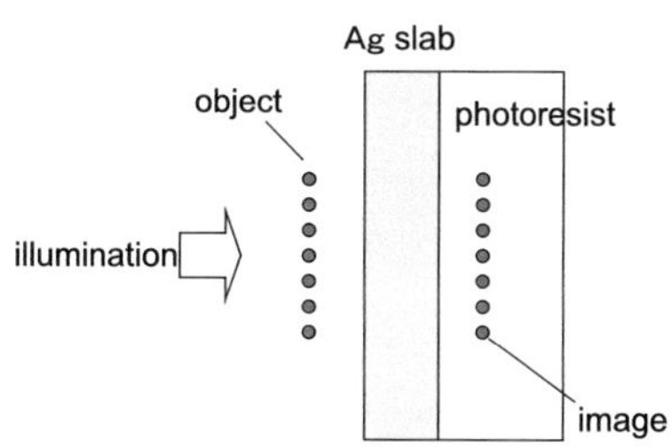

Fig. 4.8 Plasmonic super resolution [13]

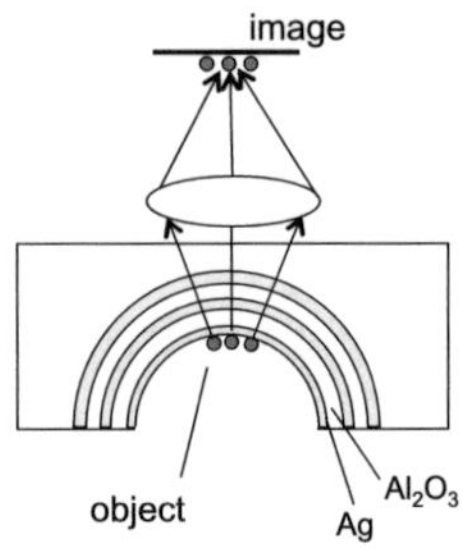

Fig. 4.9 Super resolution with magnification [14]

super resolution, we can observe the objects at a video rate. Optical microscopy can be made the objects in a liquid such as water. Hence it will be a powerful tool in biology.

4.6 Cloaking

Another interesting application of metamaterials is optical cloaking. Cloaking is a technique that makes an object invisible. One possible way is to cloak the object with a medium designed in such a way that the light detours the object, as shown in Fig. 4.10. Leonhardt proposed the spatial distribution of the refractive indexes in order to perform cloaking [17]. For this purpose, however, we need unrealistic range of refractive indexes from 0 to 36. Pendry and Smith proposed a cloaking using anisotropic metamaterials, and later they experimentally demonstrated the cloaking at the microwave frequencies [18, 19].

In the Pendry and Smith's proposal, the medium with $\mu \neq 1$ is required. This is not convenient for cloaking at optical frequencies. Shalaev et al. proposed a structure for cloaking in which metallic nanorods are distributed whose long axis is aligned along the radial direction of a cylinder, as shown in Fig. 4.11 [20]. In this method, only spatial distribution of permittivity is required. The dielectric matrix has a positive permittivity, whereas the metallic nanorod has negative one. The effective permittivity of composite materials of the nanorod and dielectric matrix can be ranged from negative to positive by changing the density of the nanorods. They designed a cylindrical composite system working as a cloaking medium. The medium with

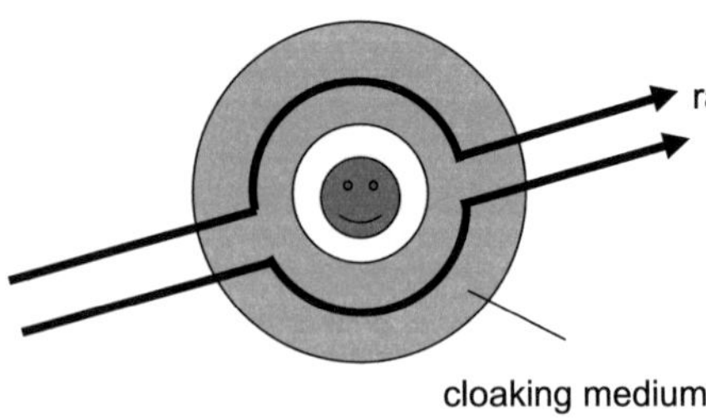

Fig. 4.10 Cloaking a object

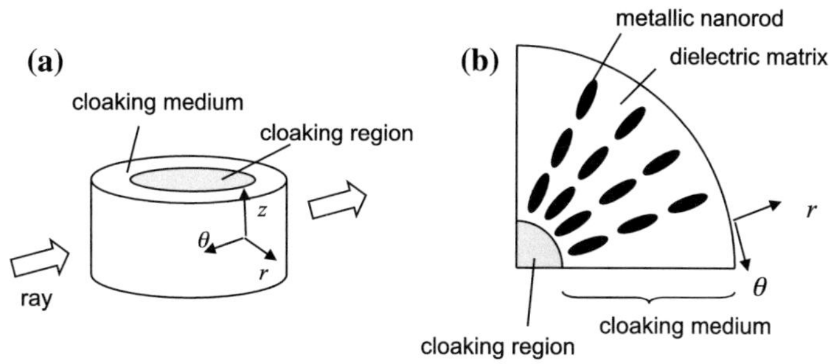

Fig. 4.11 Cloaking with nonmagnetic ($\mu = 1$) materials [20]

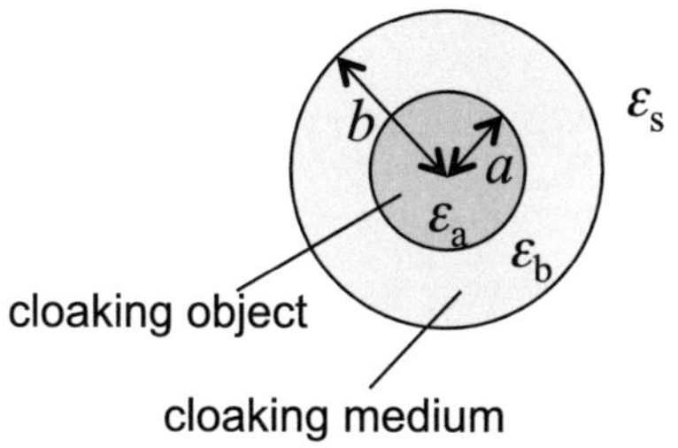

Fig. 4.12 Cloaking by core-shell structure

the effective permittivity depends on the distance from the center, since the density of the nanorod decreases. The medium with the permittivity distribution works as a cloaking medium, according to their theoretical calculation. They demonstrated the cloaking behavior by numerical simulation.

A different approach to cloak an object is proposed by Alu and Engheta [21]. They use optical response of a core-shell structure. Suppose a system that a spherical cloaking object with a diameter a and permittivity ϵ_a is surrounded by a shell with *a* thickness *b* and permittivity ϵ_b, as shown in Fig. 4.12. When the core-shell particle is much smaller than the wavelength of light, the total polarizability a can be described as [22]

$$\alpha = 4\pi b^3 \frac{(\epsilon_b - 2\epsilon_s)(\epsilon_a + 2\epsilon_b) + \left(\frac{a}{b}\right)^3 (\epsilon_a - \epsilon_b)(\epsilon_s + 2\epsilon_b)}{(\epsilon_b + 2\epsilon_s)(\epsilon_a + 2\epsilon_b) + 2\left(\frac{a}{b}\right)^3 (\epsilon_b - \epsilon_s)(\epsilon_a - \epsilon_b)} \tag{4.8}$$

where ϵ_s is the permittivity of ambient medium. When the numerator of Eq. 4.8 is zero, the polarizability is absent. This means that the particle is invisible and the core is cloaked. They calculated and showed the possibility of the cloaking at visible wavelengths, if the shell is made of a medium with negligible loss. This method is not universal, but is a possible method to realize cloaking.

References

1. V.M. Shalaev, Nat. Photonics **1**, 41 (2007)
2. F.G. Veselago, Sov. Phys. Usp **10**, 509 (1968)
3. J.B. Pendry, Phys. Rev. Lett. **85**, 3966 (2000)
4. R.A. Shelby, D.R. Smith, S. Schultz, Science **292**, 77 (2001)
5. M. Notomi, Phys. Rev. B **62**, 10696 (2000)
6. J.B. Pendry, A.J. Holden, D.J. Robbins, W.J. Stewart, IEEE Trans. Microw. Theory Tech. **47**, 2075 (1999)
7. T.J. Yen, W.J. Padilla, N. Fang, D.C. Vier, D.R. Smith, J.B. Pendry, D.N. Basov, X. Zhang, Science **303**, 1494 (2004)
8. S. Zhang, W. Fan, K.J. Malloy, S.R.J. Brueck, N.C. Panoiu, R.M. Osgood, J. Opt. Soc. Am. B **23**, 434 (2006)
9. V.M. Shalaev, W.C. Uday, K. Chettiar, H.-K. Yuan, A.K. Sarychev, V.P. Drachev, A.V. Kildishev, Opt. Lett. **30**, 3356 (2005)
10. G. Dolling, C. Enkrich, M. Wegener, C.M. Soukoulis, S. Linden, Opt. Lett. **32**, 53 (2007)
11. G. Dolling, M. Wegener, C.M. Soukoulis, S. Linden, Opt. Lett. **31**, 1800 (2006)
12. J. Valentine, S. Zhang, T. Zentgraf, E. Ulin-Avila, D.A. Genov, G. Bartal, X. Zhang, Nature **455**, 376 (2008)
13. T.Tanaka, A.Ishikawa, S.Kawata, Phys. Rev. B **73**, 125423 (2006)
14. Y. Tamayama, T. Nakanishi, K. Sugiyama, M. Kitano, Phys. Rev. B **73**, 193104 (2006)
15. N. Fang, H. Lee, C. Sun, X. Zhang, Science **308**, 534 (2005)
16. Z. Liu, H. Lee, Y. Xiong, C. Sun, X. Zhang, Science **315**, 1686 (2007)
17. U. Leonhardt, IEEE J. Sel. Top. Quantum Electron. **9**, 102 (2003)
18. J.B. Pendry, D. Schurig, D.R. Smith, Science **312**, 1780 (2006)
19. D. Schurig, J.J. Mock, B.J. Justice, S.A. Cummer, J.B. Pendry, A.F. Starr, D.R. Smith, Science **314**, 977 (2006)
20. W. Cai, U.K. Chettiar, A.V. Kildishev, V.M. Shalaev, Nat. Photonics **1**, 224 (2007)
21. A. Alu, N. Engheta, Phys. Rev. Lett. **100**, 113901 (2008)
22. C.F. Bohren, D.R. Huffman, *Absorption and Scattering of Light by Small Particles* (Wiley, New York, 1983), p. 149

Chapter 5
Spintronic Properties and Advanced Materials

Koki Takanashi and Shigemi Mizukami

Abstract Spintronics is an interactive combination of electronics and magnetics that has been growing in the twenty-first century with the development of nanotechnology. Spintronics is a new type of electronics that uses the mutual control between magnetic and other physical signals such as electrical and optical signals. As a fundamental concept for spintronics, recently, spin current has been attracting much attention. Understanding spin current means the clarification of the mechanisms underlying the mutual control of various physical signals, which is expected to result in further progress in spintronics. In this chapter, the concept of spin current and its historical background are first explained, and then advanced materials for spintronics are reviewed. Much attentions are also paid to the physical phenomena emerged from the coupling between spin in magnetic materials and electromagnetic wave. At early stage in the research of magnetism, people used electromagnetic wave as a probe to obtain various kinds of microscopic information in magnetic materials, but electromagnetic waves are used more actively in recent spintronics research. Those are not only a probe but also a tool of making spin current and of changing spin in direction and magnitudes, and those are also reviewed in this chapter.

K. Takanashi
Magnetic Materials Laboratory, Institute for Materials Research, Tohoku University,
2-1-1 Katahira, Aoba-ku, Sendai, 980-8577 , Japan
e-mail: koki@imr.tohoku.ac.jp

S. Mizukami
WPI-Advanced Institute for Materials Research, Tohoku University,
2-1-1 Katahira, Aoba-ku, Sendai, 980-8577 , Japan
e-mail: mizukami@wpi-aimr.tohoku.ac.jp

Y. Aoyagi and K. Kajikawa (eds.), *Optical Properties of Advanced Materials*,
Springer Series in Materials Science 168, DOI: 10.1007/978-3-642-33527-3_5,

5.1 Spintronics and Spin Current

5.1.1 General Introduction

Electrons have two degrees of freedom: the electric charge and the spin, which corresponds to the self-rotation angular momentum. Electric charge is the origin of electricity, and the concept of charge current, i.e., electric current, which is the flow of electric charge, was proposed in the eighteenth century. Its fundamental principles have been established through the accumulation of a huge amount of research over 200 years, leading to the dramatic development of electronics, which is indispensable in our daily lives. On the other hand, spin is the origin of magnetism. Research on magnetism has also been conducted for many years. The engineering field focused on the application of magnetism, particularly the control of magnetization, is called magnetics; like electronics, it is of industrial importance. Spintronics is an interactive combination of electronics and magnetics that has been growing in the twenty-first century with the development of nanotechnology. The magnetic properties of nanoscaled or nanostructured materials are closely correlated with other physical properties such as the electrical and optical properties. In other words, a magnetic signal may be controlled by an electrical signal, an optical signal, or any other physical signal, and vice versa. Spintronics is a new type of electronics that uses the mutual control between magnetic and other physical signals.

Spin current, i.e., the flow of spin, which corresponds to the flow of magnetism and angular momentum, has attracted much attention recently. Spin current is a fundamental concept that combines magnetic properties with other physical properties; understanding spin current is expected to clarify the mechanisms of the control between magnetic and other physical signals, leading to the further development of spintronics.

5.1.2 Concept of Spin Current

Spin is a vector; strictly speaking, spin current, the flow of spin, is a tensor. Furthermore, spin is a physical quantity that requires quantum treatment. Therefore, it is not easy to define the spin current exactly. When the direction of the spin is simplified to have two values, up or down, along the quantization axis, however, we may simply consider the spin current to be the difference in flow between spin-up and spin-down electrons, whereas the charge current is the total flow of all electrons (both spin-up and spin-down).

Here, the charge current J_{e} and spin current J_{s} are described as

$$J_{\mathrm{e}} = J_{\uparrow} + J_{\downarrow}, \tag{5.1}$$

$$J_{\mathrm{s}} = J_{\uparrow} - J_{\downarrow}, \tag{5.2}$$

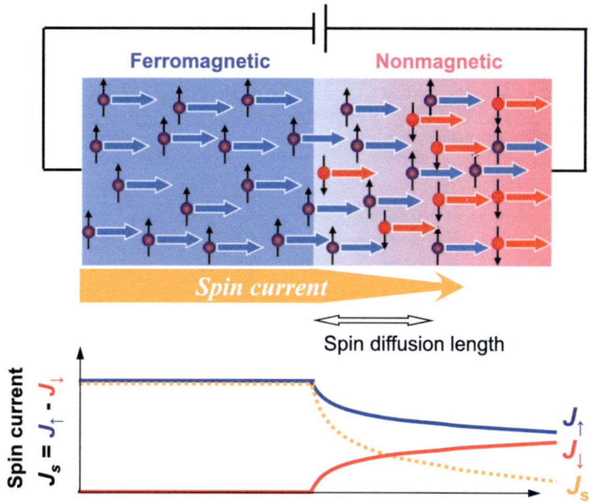

Fig. 5.1 A schematic illustration of electrical spin injection from a ferromagnetic material to a nonmagnetic material

where $J_\uparrow$ and $J_\downarrow$ are the flows of spin-up and spin-down electrons, respectively. Spin current is classified into two types: that accompanied by charge current ($J_e \neq 0$) and that not accompanied by charge current ($J_e = 0$).

Figure 5.1 schematically illustrates spin current accompanied by charge current. When charge current flows in a ferromagnetic material (strictly speaking, substances having spontaneous magnetization, including ferrimagnets), charge current as well as spin current flow because electrons in the ferromagnetic material are spin-polarized; that is, the numbers of spin-up and spin-down electrons differ. For practical use, however, it is important to extract the spin current from the ferromagnetic material. When a charge current is applied to a junction consisting of ferromagnetic and nonmagnetic materials, as shown in Fig. 5.1, spin-polarized electrons flow from the ferromagnetic material to the nonmagnetic material, and spin current can be extracted, even in the nonmagnetic material, within a certain distance from the interface. This procedure is called spin injection and is one of the most important methods of creating spin current. The spin current in the nonmagnetic material decays with the distance from the interface and finally disappears owing to spin relaxation. The range in which spin current flows in the nonmagnetic material is called the spin diffusion length and is typically on the order of 100 nm to 1 μm. Present microfabrication technology can easily handle this minute length.

Spin current that is not accompanied by charge current ($J_s \neq 0$ and $J_e = 0$) is often called pure spin current. Pure spin current is further classified into two types: that with and without electron movement. Figure 5.2a schematically illustrates pure spin current with electron movement: equal numbers of spin-up and spin-down electrons move in opposite directions. Therefore, the charge current is cancelled out ($J_e = J_\uparrow + J_\downarrow = 0$); however, spin current flows ($J_s = J_\uparrow - J_\downarrow \neq 0$). Pure

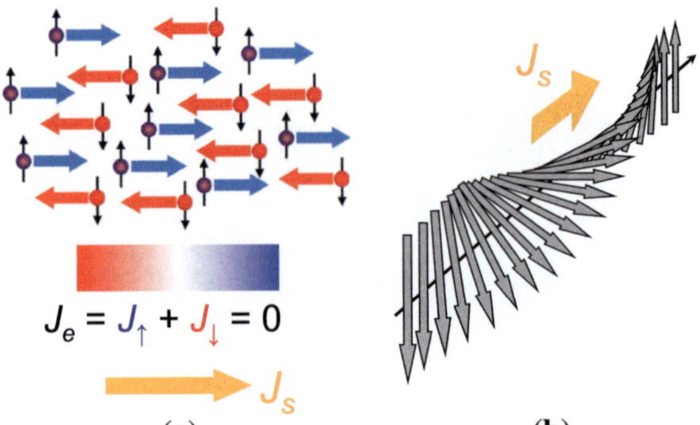

Fig. 5.2 Schematic illustrations of spin currents J_s *not* accompanied by charge current J_e: **a** pure spin current with electron movement, **b** pure spin current without electron movement

spin current is demonstrated experimentally by using nonlocal spin injection [1] (see Sect. 5.1.4) and also various methods such as the spin Hall effect [2–6] and spin pumping [7, 8]. Spin pumping is explained in Sect. 5.3.2.

A typical example of pure spin current without electron movement is a spin wave, i.e., the fluctuation of localized moments in magnetic insulators propagating spatially, as shown in Fig. 5.2b. Furthermore, circularly polarized light may also be considered a spin current (spin current of photons) in the broad sense of the term, although it is not a spin current originating from electrons.

5.1.3 Historical Background

To provide historical background on spin current research, recent developments in spintronics research are highlighted. The discovery of giant magnetoresistance (GMR) in a metallic superlattice in 1988 [9, 10] is generally considered the origin of spintronics research. A. Fert and P. Grünberg, who discovered GMR, were awarded the Nobel Prize for physics in 2007. They found that an Fe/Cr superlattice exhibited a significant decrease in resistance when the magnetic moments of the two Fe layers, which are coupled via a Cr layer, are antiparallel and become parallel to the external magnetic field.

After the discovery of GMR, research on spin-dependent transport, i.e., basic research on GMR, became very active in the field of metals. In 1995, the giant tunnel magnetoresistance (TMR) effect at room temperature was reported [11, 12]. GMR was applied practically in the read heads of hard disk drives (HDDs) in 1998, only 10 years after its discovery. During this period, a significant unique development was also observed in the semiconductor field, which was the development of the ferromagnetic semiconductors (In, Mn)As [13] and (Ga, Mn)As [14]. Research on carrier-induced magnetism also became active around this time.

In the late 1990s, people started to refer to the research field related to spin and transport in both metals and semiconductors as spintronics or spin electronics. At the same time, the direction of research changed slightly. Before this period, problems related to spin-dependent transport or the effects of magnetization on transport were the main focus of research; however, since the beginning of the twenty-first century, the effects of transport on magnetization have been attracting attention. At the same time, spin injection phenomena, as an inseparable problem, have also received attention. Examples of research topics include current-induced magnetization switching, [15, 16] self-oscillation by spin injection, [17] and domain wall motion induced by charge current [18, 19].

Regarding the history of spintronics, the following two points should be emphasized here. First, a close relationship between basic and applied research was established during the development of spintronics. The development of the GMR or TMR head for HDDs and of magnetic random access memories (MRAMs) were supported by basic research, while at the same time, such practical applications drove further developments in basic research. Second, in recent years, researchers in both the metal and semiconductor fields have come to address common research themes, such as spin-injection phenomena, the control of magnetization, and the spin Hall effect. Although the terms 'metal spintronics' and 'semiconductor spintronics' have often been used, no boundary between them currently exists in terms of research content, and the boundary between them due to the difference in materials is disappearing.

With this historical background, researchers of spintronics in the metal and semiconductor fields have begun to share a common perception. That is, in nanoscale materials, the magnetic properties are strongly correlated with other physical properties such as the electric and optical properties, and transformations between these physical properties are possible, although they are originally independent. Namely, one physical property can be controlled by another physical property, and vice versa. Spin current is the concept underlying the transformation and control of such physical properties.

5.1.4 Representative Device Structures for Spintronics

To further explain the relationship between spintronics and spin current, two representative device structures for spintronics are presented. One is the CPP (current-perpendicular-to-plane) structure, as shown in Fig. 5.3a. The basic layered structure consists of two ferromagnetic layers separated by a nonmagnetic interlayer on the nanometer scale. The electrical resistance depends on the magnetization configuration; it is usually low and high for parallel and antiparallel alignments, respectively. Previously, current was applied in the in-plane direction of the layered structure. Presently, however, the layered structure is fabricated into a pillar-type structure, and current is applied in the direction perpendicular to the plane, as shown in Fig. 5.3a. When the nonmagnetic interlayer is an insulator, the observed resistance change is TMR; when the nonmagnetic interlayer is a metal, it is GMR and is called CPP-GMR.

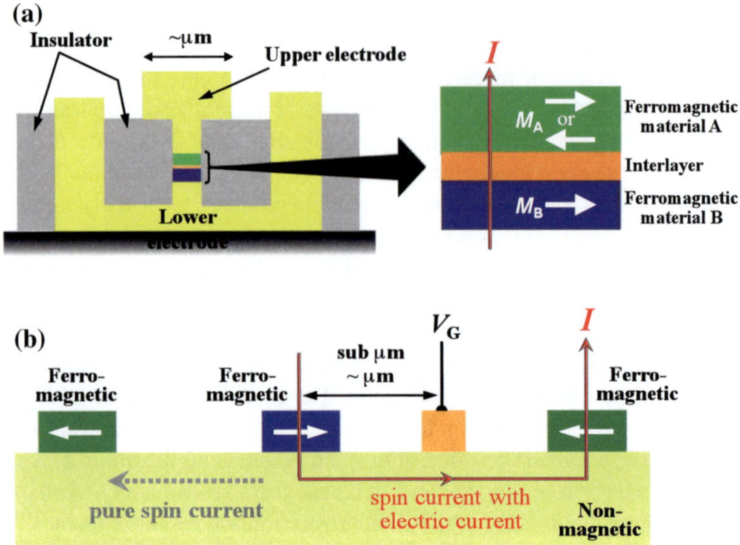

Fig. 5.3 Two representative device structures for spintronics: **a** a CPP (Current-Perpendicular-to-Plane) structure and **b** a lateral structure

The performance of this device structure is determined by the magnitude of TMR or CPP-GMR, which is roughly proportional to the spin polarizations of conduction electrons in the two ferromagnetic layers. Higher spin polarizations give rise to a higher spin current, leading in turn to higher TMR or CPP-GMR.

The other device structure is a lateral-type structure. Although the CPP structure has already been applied in HDD or MRAM technology, it is not as convenient for multi-terminal devices having higher functionalities. A typical lateral structure consisting of ferromagnetic electrodes A–C and a gate electrode on a nonmagnetic thin film is shown in Fig. 5.3b. When charge current is applied between electrodes A and B, the current can be controlled by the magnetization configuration of two ferromagnetic electrodes and also by the gate voltage between them. Next, the region in the nonmagnetic material to the left of electrode A in Fig. 5.3b is considered. In this region, no charge current flows, but spin current flows owing to spin diffusion. This is the pure spin current mentioned in Sect. 5.1.2. In other words, in this region, charge current is cancelled out ($J_e = J_\uparrow + J_\downarrow = 0$) because equal numbers of spin-up and spin-down electrons move in opposite directions; however, spin current flows ($J_s = J_\uparrow - J_\downarrow \neq 0$). The pure spin current is detected as a change in voltage, which is induced by the placement of ferromagnetic electrode C, depending on the direction of magnetization, as shown in Fig. 5.3b. The electrode configuration used to create and detect the pure spin current in this device structure is called a nonlocal geometry.

5.2 Advanced Materials for Spintronics

5.2.1 Highly Spin-Polarized Materials

In applications of magnetic materials, the most important property is usually the magnetization (specifically, the saturation magnetization M_s). For spintronics, however, the spin polarization of the conduction electrons P_c is more important than M_s. P_c depends on the difference in number between spin-up and spin-down electrons at the Fermi level E_F, in contrast with M_s, which corresponds to the difference between the total numbers of spin-up and spin-down electrons. P_c is described as

$$P_c = \frac{D_\uparrow(E_F) - D_\downarrow(E_F)}{D_\uparrow(E_F) + D_\downarrow(E_F)}, \tag{5.3}$$

where $D_{\uparrow(\downarrow)}(E_F)$ is the density of states at E_F for spin-up (down) electrons. For conventional ferromagnets such as Fe, Co, Ni, and their alloys, P_c typically ranges from 0.3 to 0.5. However, some materials are known to have $P_c = 1$; they are called half-metals. Figure 5.4 shows the schematic band structure of a half-metal. In the band structure of one type of electrons (either spin-up or spin-down), states are available at the Fermi level, and the band structure is that of a metal, whereas in the band structure of the other type of electrons, the Fermi level is in the band gap, and the band structure is that of an insulator. This is why such a material is called a half-metal.

A higher P_c increases the efficiency of spin injection, improving the performance of a spintronic device. For example, the magnitude of TMR, defined as the relative resistance change $\Delta R/R$ between the parallel and antiparallel states of the magnetization vectors in two ferromagnetic electrodes with spin polarizations P_{c1} and P_{c2}, is given by

$$\frac{\Delta R}{R} = \frac{2P_{c1}P_{c2}}{1 - P_{c1}P_{c2}}. \tag{5.4}$$

Therefore, $\Delta R/R$ becomes larger as P_{c1} and P_{c2} increase and approach 1.

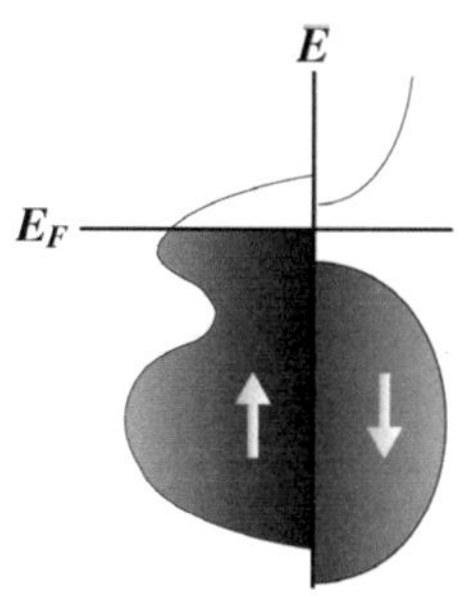

Fig. 5.4 A schematic band structure of a half metal

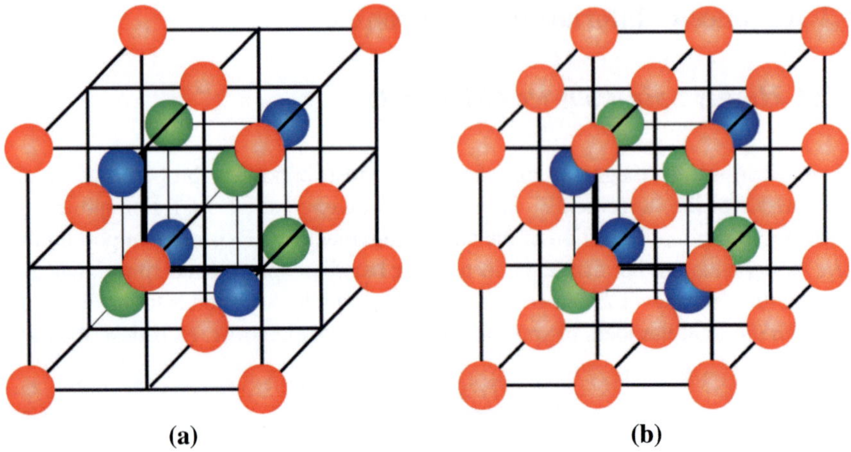

Fig. 5.5 Crystal structures of Heusler alloys: **a** Cl_b-type with the composition XYZ and **b** $L2_1$-type with the composition X_2YZ. *Red*, *blue* and *green* spheres represent X, Y and Z atoms, respectively

Research on obtaining giant TMR using half-metals has been conducted for a number of years, but it is only recently that a relatively large change in resistance has been obtained at room temperature [20–23]. Specific examples of half-metals include several Heusler alloys and oxides (such as Mn perovskite-type compounds and magnetite). However, Heusler alloys are now considered superior to oxides because of their ease of thin film preparation and high Curie temperatures. A Heusler alloy is an ordered ternary alloy. As shown in Fig. 5.5a, b, there are two types: the Cl_b type (half Heusler) with the composition XYZ and the $L2_1$ type (full Heusler) with the composition X_2YZ. All the compounds reported to date as showing large TMR have the $L2_1$-type structure.

Figure 5.6 shows the yearly change in the TMR ratio observed in Heusler alloys. The TMR ratio has increased markedly over the past several years. The largest TMR ratios ever reported are 1844 % [23] and 386 % [22] for $Co_2MnSi/MgO/Co_2MnSi$ at low temperature and $Co_2FeAl_{0.5}Si_{0.5}/MgO/Co_2FeAl_{0.5}Si_{0.5}$ at room temperature, respectively. A problem with Heusler alloys is their significant reduction in TMR at room temperature. This reduction is much more significant than that predicted from the Curie temperature. It has been suggested [24] that interface states appearing in the half-metallic band gap may contribute to the tunnelling resistance in the antiparallel state through spin mixing processes such as spin-wave excitations and inelastic scattering at room temperature. The improvement of TMR in Heusler alloys at room temperature will be an important future task.

CPP-GMR in Heusler alloys is currently attracting much attention because not only a high MR ratio but also the significantly low resistance required for device applications can be achieved. The discovery of giant TMR in a MgO tunnel junction in 2004 [25–28] was a significant breakthrough in spintronics research. Giant TMR can be obtained without using a half-metal because MgO has a spin filter effect.

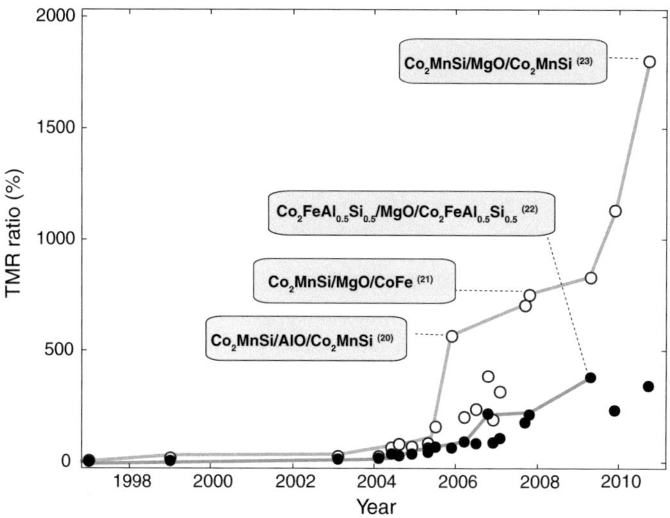

Fig. 5.6 Yearly change in the TMR ratio for Heusler alloys. Open and closed circles represent the data at a low temperature and room temperature, respectively

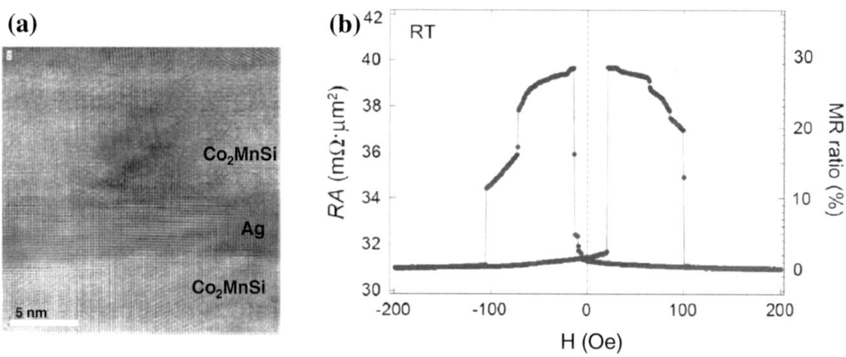

Fig. 5.7 **a** Cross-sectional high-resolution TEM images for $Co_2MnSi/Ag/Co_2MnSi$ showing epitaxial growth, and **b** the magnetoresistance (MR) curve for a $Co_2MnSi/Ag/Co_2MnSi$ CPP-GMR device [28]. *RA* is the resistance per unit area. Copyright 2009 The Japan Society of Applied Physics

In order to reduce the tunnel resistance, however, the spin filter effect decreases markedly as the thickness of the MgO barrier decreases, significantly reducing TMR. CPP-GMR shows promise for satisfying the requirements of both high MR and low resistance. Earlier, MR ratios of only a few percent were reported for CPP-GMR using conventional 3d transition metals and alloys such as CoFe. Recently, however, the use of half-metallic Heusler alloys has remarkably improved CPP-GMR. Epitaxial layered structures of $Co_2MnSi/Ag/Co_2MnSi$ were successfully prepared, and MR ratios of 30–40 % were obtained, as shown in Fig. 5.7 [28, 29]. Furthermore, an MR ratio of 75 % has been reported for $Co_2Fe_{0.4}Mn_{0.6}Si/Ag/Co_2Fe_{0.4}Mn_{0.6}Si$ [30].

5.2.2 High Magnetic Anisotropy Materials

High magnetic anisotropy is effective for the suppression of thermal fluctuations in magnetization (i.e., the superparamagnetic effect), which has become a concern with the miniaturization of devices. Given the uniaxial magnetic anisotropic constant K_u, the volume of a material V, the Boltzmann coefficient k_B, and the temperature T, the condition to be satisfied for the thermal stability of magnetization is usually described by

$$\frac{K_u V}{k_B T} > 60. \tag{5.5}$$

Therefore, the size of a device can be decreased by using a material with a larger K_u. Furthermore, high magnetic anisotropy is essential for perpendicular magnetization in a film. Perpendicular magnetization can be achieved when K_u exceeds the shape magnetic anisotropy $\mu_0 M_s^2/2$, i.e., when

$$K_u > \mu_0 M_s^2/2 \tag{5.6}$$

is satisfied.

Several types of material are used for perpendicular magnetization: Co-based granular alloys such as Co(Pt)-SiO_2, rare earth-transition metal amorphous alloys such as TbFeCo, metallic superlattices such as Ni/Co and Co/Pt, and $L1_0$-ordered alloys such as FePt, CoPt, and FePd. Among them, $L1_0$-ordered alloys in particular are known to show very high magnetic anisotropy. The $L1_0$-ordered alloys have the crystal structure shown in Fig. 5.8. The [001] (c-axis) direction is usually the easy magnetization axis. The K_u value of $L1_0$-FePt is the highest among the $L1_0$-ordered alloys, that is, 7×10^6 J/m^3 higher than that of the permanent magnet $Nd_2Fe_{14}B$.

Perpendicular magnetization has attracted much attention not only for HDD technology, but also for spintronic devices such as MRAM, because of several advantages

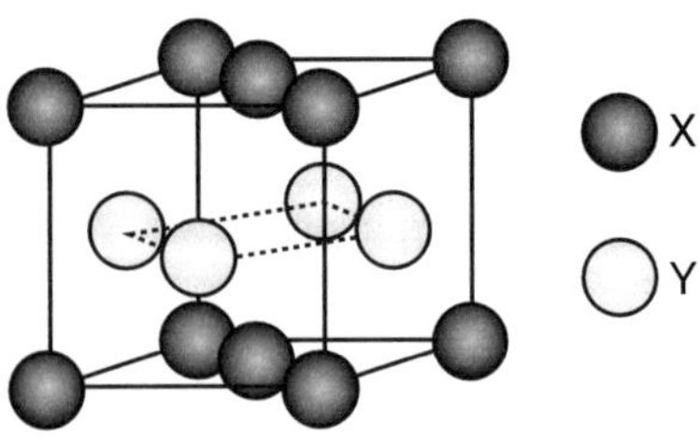

Fig. 5.8 Crystal structure of $L1_0$-ordered alloy with the composition XY

it offers for high-density integration, such as a reduced critical current density for magnetization switching and the ability to have memory cells with an arbitrary aspect ratio. Current-induced magnetization switching in perpendicularly magnetized films was first demonstrated in 2006 using metallic superlattices [31, 32] and $L1_0$-FePt [33]. Subsequently, perpendicular magnetization has been crucial for high-density MRAM [34, 35]. Perpendicularly magnetized films can generally be used as a perpendicular spin polarizer (i.e., injector and/or detector). As a representative example, giant spin Hall effects were discovered in Au thin films, using a $L1_0$-FePt(001) film as a perpendicular spin polarizer [36, 37].

5.2.3 Semiconductors for Spintronics

High-efficiency electrical spin injection from a ferromagnet into a semiconductor is important for the development of spintronics [38]. However, it is not easy because of the problem explained below. The spin injection efficiency (P_N) is defined as the ratio of spin current (J_s) to charge current (J_e) at the interface between ferromagnetic and nonmagnetic materials, as shown in Fig. 5.1:

$$P_\mathrm{N} = \left.\frac{J_\mathrm{s}}{J_\mathrm{e}}\right|_\mathrm{interface} = \left.\frac{J_\uparrow - J_\downarrow}{J_\uparrow + J_\downarrow}\right|_\mathrm{interface}. \tag{5.7}$$

Then,

$$P_\mathrm{N} = \frac{P_\mathrm{F}}{1 + (R_\mathrm{N}/R_\mathrm{F})(1 - P_\mathrm{F}^2)} \tag{5.8}$$

is obtained from a calculation using a simple model, [39] where P_F, R_F, and R_N are the spin polarization ratio of the ferromagnetic material and the electric resistances of the ferromagnetic and nonmagnetic materials, respectively. Therefore, when the electric resistance of the nonmagnetic material is much larger than that of the ferromagnetic material, i.e., $R_\mathrm{N} \gg R_\mathrm{F}$, then $P_\mathrm{N} \ll P_\mathrm{F}$ holds; in other words, the spin polarization ratio decreases significantly at the interface. As a result, negligible spin current flows into the nonmagnetic material. That is, when the ferromagnetic material is a metal with a low resistance and the nonmagnetic material is a semiconductor with a high resistance, spin injection is extremely difficult. An effective solution to this resistance mismatch problem is the introduction of a large contact resistance between the metal and the semiconductor, which may be realized by the formation of a Schottky barrier or the insertion of an additional tunnel barrier.

On the basis of electrical detection of spin signals, several groups [40–49] have reported successful electrical spin injection from a ferromagnetic metal to a semiconductor, including Si or a compound such as GaAs, by using a tunnel barrier at the interfaces. Furthermore, spin pumping was recently shown to enable efficient spin injection even through low-resistance interfaces, i.e., ohmic contacts [50].

Another approach to efficient electrical spin injection into a semiconductor is to also use a ferromagnetic semiconductor as a spin injection source. This is the most straightforward solution to the resistance mismatch problem. However, for practical applications, i.e., spin injection at room temperature, the Curie temperatures of many ferromagnetic semiconductors are below room temperature. The synthesis of ferromagnetic semiconductors having a high Curie temperature is a major challenge in the field of spintronics. Significant developments in enhancing the Curie temperature were recently seen in metal oxide systems with magnetic dopants [51].

5.2.4 Other Topics

Magnetic insulators such as $Y_3Fe_5O_{12}$ (usually abbreviated as YIG) have been studied extensively for various applications to magneto-optical devices, microwave devices, and memories; they are regarded as rather classical magnetic materials. Recently, however, they have been attracting much attention again as advanced materials for spintronics. In a magnetic insulator, charge current does not flow, but spin current does. A spin wave (i.e., magnon) is a carrier for the spin current. Spin-wave spin current in a magnetic insulator may be generated by different methods, for example, spin pumping [52] and the spin Seebeck effect [53]. Furthermore, in a junction system consisting of Pt/YIG/Pt, [52] it was clearly demonstrated that spin current with electron movement, as shown in Fig. 5.2a, can be converted to that without electron movement, i.e., spin-wave spin current, as shown in Fig. 5.2b, and vice versa. Spin-wave spin current in a magnetic insulator may flow over a relatively long distance on a macroscopic scale because the magnetization damping is extremely small, and the spin-wave decay length can be on the order of a centimeter.

Organic molecules, including nano-carbonaceous molecules such as graphene, carbon nanotubes, and fullerene, have also attracted much attention as advanced materials for spintronics [54]. Spin-orbit interaction is very weak in these materials, which is expected to lead to a relatively long spin relaxation time and spin diffusion length. Furthermore, novel behavior for spin transport unlike that observed in inorganic materials may appear.

5.3 Optical Properties Related to Spin Dynamics

5.3.1 Overview

As stated in the last section, magnetic materials can be used in spin transport devices because electron charge couples strongly to spin angular momentum in ferromagnets. Researchers in spintronics have also explored physical phenomena emerging from the coupling between spin and electromagnetic waves in magnetic materials. At an early stage in research on magnetism, people used electromagnetic waves as a probe to obtain various types of microscopic information regarding magnetic materials, but electromagnetic waves are now used more actively in spintronics research. They are not only a probe but also a tool for creating spin current and changing the direction and magnitude of magnetization as if the electromagnetic waves are an electric current generator or static magnetic field.

The interaction between electron spin and the magnetic field of electromagnetic waves is much weaker than that between electron charge and the electric field of electromagnetic waves. However, radiation energy can be efficiently transferred to the kinetic motion of spin when the radiation frequency is identical to the spin resonance frequency, that is, in electron spin resonance (ESR) and/or ferromagnetic resonance (FMR), which were discovered in the last century. In ESR or FMR, non-equilibrium electron spins, or magnons, can be created by absorption of electromagnetic radiation with a frequency in the GHz range under an external magnetic field, so one can manipulate the spin angular momentum by applying not current but radiation in experiments. ESR and FMR have been used as tools to investigate various properties related to spin, such as Landé's g-tensor, the spin relaxation time in paramagnets, [55] spin-wave stiffness, magnetic anisotropy, and Gilbert damping in ferromagnets or ferrimagnets [56]. The research field related to FMR is becoming increasingly broad. FMR experiments can be done not only in macroscopic thin films but also in the nanoscale regime. One of the most important findings regarding FMR is that it can be used as a spin current generation technique, so-called spin pumping, [57] which is reviewed in Sect. 5.3.2.

The interaction of electron charge and the electric field of electromagnetic waves is much stronger than that in FMR and ESR. However, the electric field of electromagnetic waves has little direct effect on spin motion. Instead, it can couple to the spin indirectly with the aid of spin-orbit interaction. The classic example is the so-called *magneto-optical effects*, such as the Faraday, Kerr, and Voigt effects, which have been studied extensively since their discovery [58–61]. They are used to detect the spin or magnetization direction not only in magnetic materials but also in semiconductors. Visible lights or lasers, which were used as probes at an early stage of the research, are now also used as tools for the creation and manipulation of spin or magnetization. A classic but important light-induced spin phenomenon is optical orientation [62]. Non-equilibrium spin polarization can be created by a circularly polarized light in the case of III–V semiconductor compounds, in which point-inversion asymmetry and spin-orbit interaction play key roles. Much progress related to optical orientation has been made in III–V-based semiconductors, i.e. not only for electron spin but also for nuclear spin manipulation, which is related to quantum computing [63].

In the last two decades, significant progress was also made in this field by using experimental techniques involving ultrashort pulse lasers. The so-called *ultrafast demagnetization* observed in nickel films was one of the triggers of this research [64]. Subsequently, an all-optical probe of coherent spin precessional dynamics has also been reported [65]. A highlight of this type of research is *all-optical magnetization reversal* on a very short time scale [66]. These ultrafast optical-induced spin dynamics can also be seen in magnetic and nonmagnetic semiconductors and insulating oxides, and this is one of the most exciting issues in spintronics [67]. We focus on all-optical studies only in metallic magnets in Sect. 5.3.3.

5.3.2 Microwave Technique and Spin Pumping

The low-energy magnetization dynamics in magnetic materials are described well by the Landau-Lifshitz-Gilbert equation [68]:

$$\frac{d\boldsymbol{m}}{dt} = -\gamma \boldsymbol{m} \times \boldsymbol{H}_{\mathrm{eff}} + \alpha \boldsymbol{m} \times \frac{d\boldsymbol{m}}{dt}. \tag{5.9}$$

Here, $\boldsymbol{m}$ is the unit vector of magnetization; γ is the gyromagnetic ratio; $\boldsymbol{H}_{\mathrm{eff}}$ is the effective magnetic field, including the relevant magnetic anisotropy field and external magnetic field; and α is the Gilbert damping constant. The first and second terms express the precession and relaxation torques acting on the magnetization, respectively, as shown in Fig. 5.9a, where we assume a magnetic film having a magnetically easy axis parallel to the film normal for simplicity. When a circularly polarized microwave with a frequency identical to the FMR frequency is incident to the film along the film normal (Fig. 5.9b), the microwave magnetic field $\boldsymbol{h}$ excites continuous precession of the magnetization because the microwave magnetic field torque is balanced by the damping torque.

If a conducting material layer is in contact with a magnetic film, magnetization precession can generate spin angular momentum current at the interface essentially without a charge current; that is, spin pumping can occur (Fig. 5.10). The adiabatic spin pumping theory yields the following expression for the spin-pumping-induced spin current: [7, 57, 69]

$$\boldsymbol{J}_{\mathrm{s}} = \frac{\hbar}{4\pi} g_{\mathrm{mix}} \boldsymbol{m} \times \frac{d\boldsymbol{m}}{dt}. \tag{5.10}$$

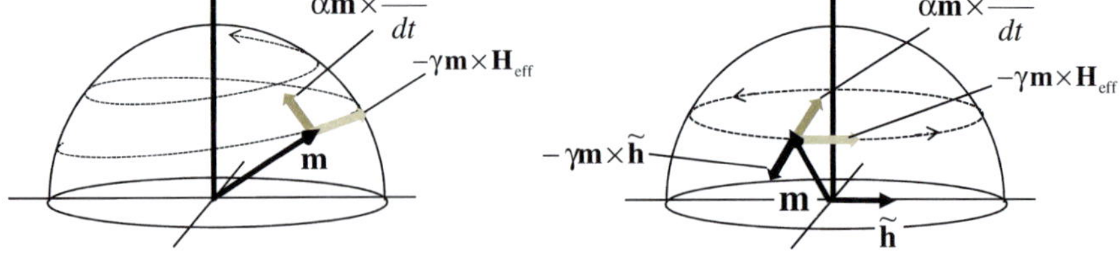

Fig. 5.9 **a** Two torques proportional to the effective magnetic field $\boldsymbol{H}_{\mathrm{eff}}$ or Gilbert damping constant α, which drive the decayed precessional dynamics of magnetization $\boldsymbol{m}$ (broken curve). **b** Stationary precession of magnetization generated by microwave absorption, where the damping torque is balanced by the torque of the microwave magnetic field $\boldsymbol{h}$

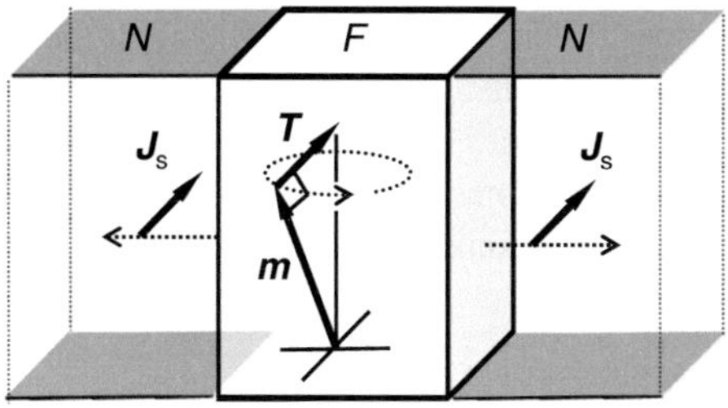

Fig. 5.10 Schematic of spin pumping. Spin current $\boldsymbol{J}_S$ is generated by fast precession of magnetization $\boldsymbol{m}$ in magnetic layer F into nonmagnetic conductive layer N. Spin direction of current is parallel to damping torque owing to spin pumping $\boldsymbol{T}$, which is parallel to the Gilbert damping torque

Here, g_{mix} is the spin mixing conductance, which is the conductance for spin current perpendicular to the magnetization direction. It is, in principle, different from the conductance for a current consisting of the longitudinal component of the spin, but both are roughly the same order of magnitude at a metallic interface [57]. The vector of the spin current in (5.10) is parallel to the damping torque (Fig. 5.10), so both RF and DC spin current can be generated during magnetization precession, depending on the precession amplitude (cone angle).

The spin pumping effect can be observed as an amplification of conduction electron spin resonance (CESR) signals [70]. More conveniently, it may be measured as enhanced Gilbert damping in very thin magnetic films [8, 71, 72]. Emission of spin current from a magnetic film causes a reaction; this reaction can be expressed as additional torque [7, 57, 69]:

$$\boldsymbol{T} = \frac{\gamma}{M_S d} \frac{\hbar}{4\pi} g_{\text{mix}} \boldsymbol{m} \times \frac{d\boldsymbol{m}}{dt}. \tag{5.11}$$

Here, d and M_s are the magnetic layer thickness and saturation magnetization, respectively. $\boldsymbol{T}$ is added to the left-hand side of (5.9) if spin pumping is operative (Fig. 5.10). This reaction torque is a damping-type torque similar to the Gilbert damping torque in (5.9), implying that it describes a loss of spin angular momentum and energy in a magnetic film resulting from spin current emission. The Gilbert damping constant enhanced by spin pumping can be written as

$$\alpha' = \alpha + \frac{\gamma \hbar g_{\text{mix}}}{4\pi M_S d}. \tag{5.12}$$

If the magnetic film is sufficiently thin, the damping constant should increase as a function of the inverse of the thickness of the magnetic layer; this was observed in FMR experiments in very thin magnetic films [8, 71, 72].

Recently, E. Saitoh and his colleagues discovered the so-called *inverse spin Hall effect* and developed a method of electrical detection of spin current injected into metals by spin pumping [5]. They extended this technique to other combinations of materials, i.e., ferromagnetic metals and p- or n-type semiconductors, [50, 73] a ferrimagnetic insulator and a metal, [52] and a ferrimagnetic insulator and an oxide semiconductor [74]. They also proposed an expression for the mixing conductance at a ferromagnetic-semiconductor contact that is somewhat different from the original one for a metallic contact [57].

5.3.3 Optical Techniques and Ultrafast Spin Dynamics

(1) **Ultrafast change in magnetic properties** Magnetic materials have three degrees of freedom: the electron, spin, and lattice (Fig. 5.11a). The temperatures for each subsystem are identical under thermal equilibrium at a relatively long time scale, so magnetization motion can be properly described by (5.9) at a temperature T. This is a type of adiabatic approximation in which the magnetization motion can be considered while neglecting the electron, spin, and phonon systems. However, it is far from thermal equilibrium when the material is suddenly irradiated with an intense laser pulse. A number of non-equilibrium electron-hole pairs, Stoner excitations, and/or magnons and phonons are created in each subsystem; the subsystems can each be characterized by a distinctive time-dependent temperature, and their dynamics are described by the so-called three-temperature model [64, 75]:

$$C_{\mathrm{E}}\frac{dT_{\mathrm{E}}}{dt} = -G_{\mathrm{ES}}(T_{\mathrm{E}} - T_{\mathrm{S}}) - G_{\mathrm{EL}}(T_{\mathrm{E}} - T_{\mathrm{L}}) + P_{\mathrm{laser}}(t), \tag{5.13}$$

$$C_{\mathrm{S}}\frac{dT_{\mathrm{S}}}{dt} = -G_{\mathrm{ES}}(T_{\mathrm{S}} - T_{\mathrm{E}}) - G_{\mathrm{SL}}(T_{\mathrm{S}} - T_{\mathrm{L}}), \tag{5.14}$$

$$C_{\mathrm{L}}\frac{dT_{\mathrm{L}}}{dt} = -G_{\mathrm{EL}}(T_{\mathrm{L}} - T_{\mathrm{E}}) - G_{\mathrm{SL}}(T_{\mathrm{L}} - T_{\mathrm{S}}). \tag{5.15}$$

Here, $T_{\mathrm{E}}(C_{\mathrm{E}})$, $T_{\mathrm{S}}(C_{\mathrm{S}})$, and $T_{\mathrm{L}}(C_{\mathrm{L}})$ are the system temperature (specific heat) of the electron, spin, and lattice system, respectively. G_{EL} is the energy transfer rate between the electron and the lattice system, G_{ES} is that between the electron and the spin system, and G_{SL} is that between the spin and the lattice system. A typical temperature variation against time is shown schematically in Fig. 5.11b and can be calculated from these equations with a proper function for the absorbed laser power P_{laser} in (Fig. 5.13a). Initially, the electrons absorb the light pulse, and then the electron temperature increases suddenly (within a few tens of fs) owing to electron-electron scattering. The electronic specific heat is an order smaller than that for the lattice, so an electron temperature of several hundred K corresponds to a lattice temperature of a few tens of K. Subsequently, the energy is transferred to the spin and/or lattice system via electron-spin or electron-phonon scattering, respectively. Energy is also transferred from the spin to the lattice system via spin-lattice relaxation, and ultimately all the subsystems reach thermal equilibrium. This process of energy flow is denoted in Fig. 5.11a by arrows. A sudden change in spin

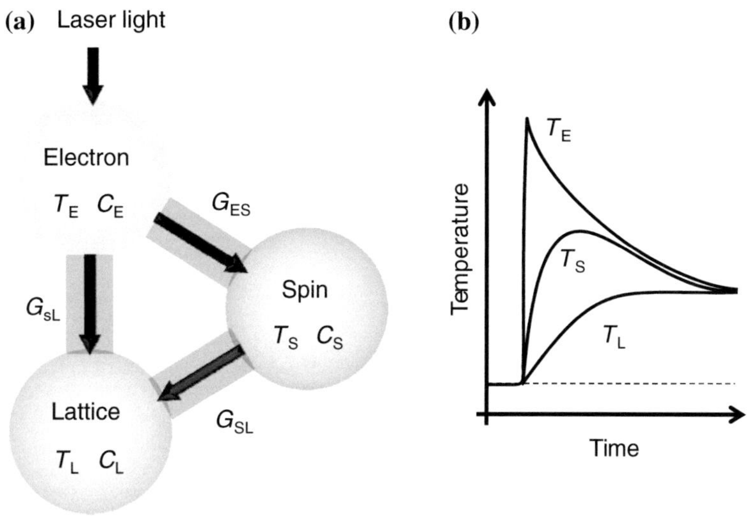

Fig. 5.11 Schematic of **a** three-temperature model used in ultrafast magneto-optical response in magnetic conductors and **b** temperature in each system as a function of time [64, 75]

temperature can be monitored as an ultrafast change in the magneto-optical Kerr effect signal, which indicates ultrafast demagnetization. This ultrafast change in the magnetization magnitude has also been proven by the detection of THz emission due to a rapid change in the induction [76] or can be recognized as the subsequent effective pulse field that induces precessional dynamics, [65] as described in the next section.

The reason for the very rapid decrease in magnetization, characterized by the demagnetization time τ_M (i.e., the mechanism for the electron-spin scattering process G_{ES}) is still under investigation. One group has suggested that it is mediated by spin-flip scattering of phonons with spin-orbit interaction [77]. Another group has studied the ultrafast demagnetization process in various magnetic materials, including advanced materials with high spin polarization, and suggested that the demagnetization time correlates to the degree of spin polarization of electrons at the Fermi level [78].

Another interesting issue is the ultrafast quenching of magnetization under a high pump fluence, which is intuitively expected if one can increase the spin temperature near or beyond the magnetic ordering temperature T_C. However, this has never been observed yet.

Beyond the original three-temperature model, advanced models that properly describe the spin dynamics have been investigated. One is the atomistic Landau-Lifshitz-Gilbert or Landau-Lifshitz-Bloch equation, which takes account of the stochastic field [79, 80]. Another is the microscopic three-temperature model, an extension of the original three-temperature model that considers the spin-flip process due to spin-orbit interaction [81]. Recent theoretical work suggested a partial equivalence

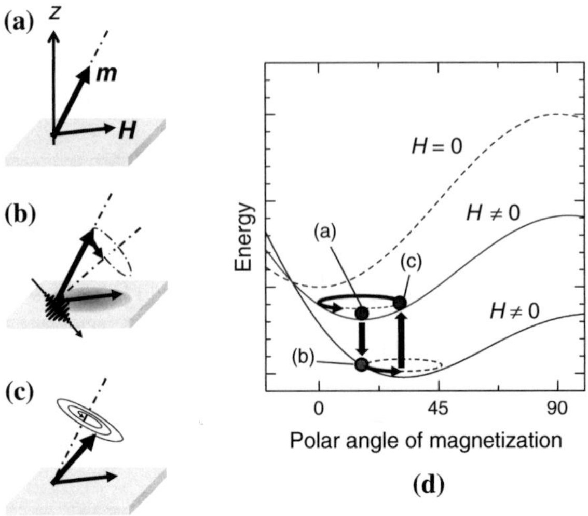

Fig. 5.12 Thermal mechanism of excitation of magnetization precession in a perpendicularly magnetized film by pulsed laser. **a** Initial state. Magnetization ***m*** is tilted by applying magnetic field ***H***, **b** film irradiated by pulsed laser, **c** after heat dissipation. **d** Change in energy of magnetization and corresponding variation in magnetization angle

between the Landau-Lifshitz-Bloch approach and the microscopic three-temperature model, [82] and exploration of a unifying physical picture continues.

(2) **Fast magnetization precession and relaxation** The ultrafast increase in spin temperature decreases not only the shape magnetic anisotropy (via a reduction in the saturation magnetization) but also the crystalline magnetic anisotropy, because the crystalline magnetic anisotropy constant is very sensitive to temperature. The rapid change in the magnitude of the magnetic anisotropy can drive coherent magnetization precession [65]. Let us imagine a perpendicularly magnetized magnetic film with a large perpendicular uniaxial anisotropy in which the magnetization is tilted by applying a magnetic field, as shown in Fig. 5.12a. The spin temperature rises and the anisotropy decreases as soon as the laser pulse hits the film, so the magnetization begins to move toward a new equilibrium direction (Fig. 5.12b). However, heat dissipation is more rapid than magnetization motion; thus, the magnetization precesses about the initial equilibrium direction (Fig. 5.12c). This process can also be understood in terms of energy considerations, as shown in Fig. 5.12d. The ultrafast temperature increase reduces the free energy profile for magnetization; consequently, the magnetization is in an excited state, which induces magnetization precession.

Although the mechanism behind the ultrafast demagnetization process is still open, this precession effect can be used as a tool to investigate ultrafast precession of spins in some materials. Spin precession is induced only by a laser light pulse, and no coils or inductive circuit elements, which are required to generate a pulsed magnetic field in other methods, are involved. Thus, one can achieve the ultimate time resolution (better than hundreds of fs, corresponding to THz) in this way. The

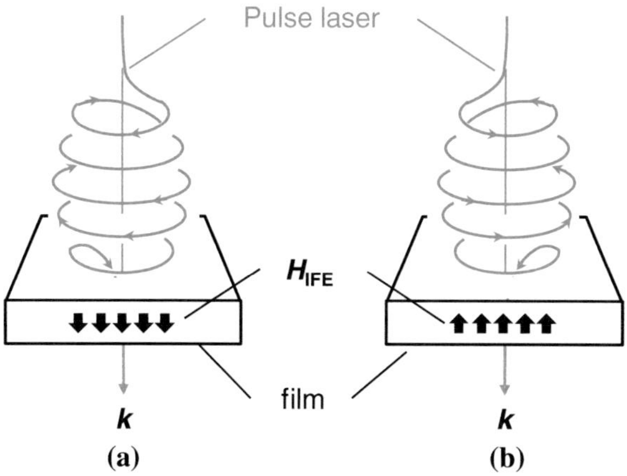

Fig. 5.13 Schematic of the inverse Faraday effect (IFE) due to an ultrafast laser pulse in a film. Effective magnetic field $\boldsymbol{H}_{\mathrm{IFE}}$ is created by a circularly polarized pulsed laser with a propagation vector $\boldsymbol{k}$. The direction of $\boldsymbol{H}_{\mathrm{IFE}}$ is reversed when polarization is in opposite directions: the left-handed (**a**) and the right-handed case (**b**) [88]

magnitude of the created effective field pulse is proportional to the pump pulse fluence and the original magnetic anisotropy field. The magnetic anisotropy field can exceed several tens of kOe for materials with high magnetic anisotropy; consequently, an effective field pulse comparable to several kOe can easily be generated. These values may not be easily accessible in conventional microwave techniques but are required for investigating the exchange mode of ferrimagnets [83] and magnets with high magnetic anisotropy [84].

(3) **Ultrafast magnetization reversal** Ultrafast demagnetization of most magnetic metals can be understood as a thermal effect because it does not depend on the helicity of the polarization vector of the pump laser pulse [85]. On the other hand, the change in the magnitude of the magnetization depends on the helicity of the circularly polarized light pulse in some oxides and semiconductors. This indicates that quantum mechanical interaction between light and spin is present, the so-called the *inverse-Faraday effect* (IFE) [86, 87]. When circularly polarized light passes through a medium, it induces magnetization in the medium. One can also think that magnetization can be induced by an effective magnetic field due to the IFE, $\boldsymbol{H}_{\mathrm{IFE}}$. In isotropic media, $\boldsymbol{H}_{\mathrm{IFE}}$ can be expressed as

$$\boldsymbol{H}_{\mathrm{IFE}} = \alpha_{\mathrm{MO}}[\boldsymbol{E}(\omega) \times \boldsymbol{E}(\omega)^*], \tag{5.16}$$

where α_{MO} is the magneto-optic susceptibility for the medium, and $\boldsymbol{E}$ is the electric field vector of the light [87]. For circularly polarized light, the created effective field is parallel to the propagation direction of light $\boldsymbol{k}$, and the field direction can be changed by varying the rotation direction of the polarization vector. This effect is larger in materials exhibiting a larger Faraday effect because both are characterized by α_{MO}, and the size of this effect is proportional to the light intensity (Fig. 5.13).

A. V. Kimel et al. have reported, for the first time, ultrafast spin precession in $DyFeO_3$ induced by the IFE using a fs pulse laser [88]. Subsequently, C. D. Stanciu et al. demonstrated full magnetization reversal for a short period of time in perpendicularly magnetized GdFeCo alloy thin films using a single-shot laser pulse with a duration of 40 fs [66]. The direction of magnetization in the film could be reversed arbitrarily by changing the helicity (i.e., left-handed or right-handed) of the polarization of the light pulse (Fig. 5.13). These experimental results were explained well by a combination of ultrafast demagnetization and the IFE using the atomistic Landau-Lifshitz-Bloch equation [89]. Very interestingly, purely thermal ultrafast magnetization reversal (that is, without the IFE) has also been reported quite recently [90]. These ultrafast magnetization reversal effects will be a key technology for next-generation high-density magnetic storage applications.

References

1. F.J. Jedema et al., Nature **410**, 345 (2001)
2. Y.K. Kato et al., Science **306**, 1910 (2004)
3. J. Wunderlich et al., Phys. Rev. Lett. **94**, 047204 (2005)
4. S.O. Valenzuela, M. Tinkham, Nature **442**, 176 (2006)
5. E. Saitoh et al., Appl. Phys. Lett. **88**, 182509 (2006)
6. T. Kimura et al., Phys. Rev. Lett. **98**, 156601 (2007)
7. Y. Tserkovnyak et al., Phys. Rev. B **66**, 224403 (2002)
8. S. Mizukami et al., Phys. Rev. B **66**, 104413 (2002)
9. M.N. Baibich et al., Phys. Rev. Lett. **61**, 2472 (1988)
10. G. Binasch et al., Phys. Rev. B **39**, 4828 (1989)
11. T. Miyazaki, N. Tezuka, J. Magn. Magn. Mater. **139**, L231 (1995)
12. J.S. Moodera et al., Phys. Rev. Lett. **74**, 3273 (1995)
13. H. Ohno et al., Phys. Rev. Lett. **68**, 2664 (1992)
14. H. Ohno et al., Appl. Phys. Lett. **69**, 363 (1996)
15. E.B. Myers et al., Science **285**, 867 (1999)
16. D. Chiba et al., Phys. Rev. Lett. **93**, 216602 (2004)
17. S.I. Kiselev et al., Nature **425**, 380 (2003)
18. A. Yamaguchi et al., Phys. Rev. Lett. **92**, 077205 (2004)
19. M. Yamanouchi et al., Nature **428**, 539 (2004)
20. K. Inomata et al., Jpn. J. Appl. Phys. **42**, L419 (2003)
21. Y. Sakuraba et al., Appl. Phys. Lett. **88**, 192508 (2006)
22. N. Tezuka et al., Appl. Phys. Lett. **94**, 162504 (2009)
23. T. Taira et al., in *Proceedings of 55th Annual Conference on Magnetism and Magnetic Materials, Abstracts*, Atlanta, Georgia, USA, 14–18 Nov 2010, pp. 118–119, BH-10
24. P. Mavropoulos et al., Phys. Rev. B **72**, 174428 (2005)
25. S. Yuasa et al., Jpn. J. Appl. Phys. **43**, L588 (2004)
26. S. Yuasa et al., Nat. Mater. **3**, 868 (2004)
27. S.S.P. Parkin et al., Nat. Mater. **3**, 862 (2004)
28. T. Iwase et al., Appl. Phys. Exp. **2**, 063003 (2009)
29. Y. Sakuraba et al., Phys. Rev. B **82**, 094444 (2010)
30. J. Sato et al., Appl. Phys. Exp. **4**, 113005 (2011)

31. S. Mangin et al., Nat. Mater. **5**, 210 (2006)
32. H. Meng, J.-P. Wang, Appl. Phys. Lett. **88**, 172506 (2006)
33. T. Seki et al., Appl. Phys. Lett. **88**, 172504 (2006)
34. H. Yoda et al., Curr. Appl. Phys. **10**, e87 (2010)
35. S. Ikeda et al., Nat. Mater. **9**, 721 (2010)
36. T. Seki et al., Nat. Mater. **7**, 125 (2008)
37. B. Gu et al., Phys. Rev. Lett. **105**, 216401 (2010)
38. T. Taniyama et al., NPG Asia Mater. **3**, 65 (2011)
39. M. Ziese, in *Spin Electronics*, ed. by M. Ziese, M.J. Thornton (Springer, Berlin, 2001) p. 396
40. B.T. Jonker et al., Nat. Phys. **3**, 542 (2007)
41. X. Lou et al., Nat. Phys. **3**, 197 (2007)
42. I. Appelbaum et al., Nature **447**, 295 (2007)
43. O.M.J. van' t Erve et al., Appl. Phys. Lett. **91**, 212109 (2007)
44. S. Dash et al., Nature **462**, 491 (2009)
45. T. Inouchi et al., Appl. Phys. Exp. **2**, 023006 (2009)
46. Y. Ando et al., Appl. Phys. Lett. **94**, 182105 (2009)
47. T. Sasaki et al., IEEE Trans. Magn. **46**, 1436 (2010)
48. H. Saito et al., Appl. Phys. Lett. **96**, 012501 (2010)
49. T. Uemura et al., Appl. Phys. Lett. **99**, 082108 (2011)
50. K. Ando et al., Nat. Mater. **10**, 655 (2011)
51. S.B. Ogale, Adv. Mater. **22**, 3125 (2010), and references therein
52. Y. Kajiwara et al., Nature **464**, 262 (2010)
53. K. Uchida et al., Nat. Mater. **9**, 894 (2010)
54. M. Shiraishi, T. Ikoma, Physica E **43**, 1295 (2011)
55. C.P. Poole Jr, *Electron Spin Resonance* (Dover, Mineola, New York, 1996)
56. Z. Frait, D. Fraitova, Spinwave resonance in metals, in *Spin Waves and Magnetic Excitations*, ed. by A.S. Borovik-Romanov, S.K. Sinha (North-Holland, Amsterdam, 1988), Pt. 2, Chap. 1, p. 1
57. Y. Tserkovnyak et al., Phys. Rev. Lett. **88**, 117601 (2002)
58. M. Faraday, Phil. Trans. Roy. Soc. **136**, 1 (1846)
59. J. Kerr, Rept. Brit. Assoc. Adv. Sci. **40**, (1876)
60. J. Kerr, Phil. Mag. **3**, 321 (1877)
61. W. Voigt, Nachricht Gesellschaft Wiss. Goettingen II. Math.-Phys. Kl. **4**, 355 (1898)
62. F. Meier, B.P. Zakharchenya (ed.), *Optical Orientation* (Elsevier, Amsterdam, 1984)
63. D.D. Awschalom, D. Loss, N. Samarth (ed.), *Semiconductor Spintronics and Quantum Computing* (Springer, Germany, 2002)
64. E. Beaurepaire et al., Phys. Rev. Lett. **76**, 4250 (1996)
65. M. van Kampen et al., Phys. Rev. Lett. **88**, 227201 (2002)
66. C.D. Stanciu et al., Phys. Rev. Lett. **99**, 047601 (2007)
67. A. Kirilyuk et al., Rev. Mod. Phys. **82**, (2010)
68. T.L. Gilbert, IEEE Trans. Magn. **40**, 3443 (2004)
69. Y. Tserkovnyak et al., Rev. Mod. Phys. **77**, 1375 (2005)
70. R.H. Silsbee et al., Phys. Rev. B **19**, 4382 (1979)
71. S. Mizukami et al., Jpn. J. Appl. Phys. **40**, 580 (2001)
72. R. Urban et al., Phys. Rev. Lett. **87**, 217204 (2001)
73. K. Ando, E. Saitoh: Nat. Commun. **3**, 629 (2012). doi:10.1038/ncomms1640
74. Z. Qiu et al., Appl. Phys. Lett. **100**, 022402 (2012)
75. B. Koopmans, Time-resolved Kerr-effect and spin dynamics in itinerant ferromagnets, in *Handbook of Magnetism and Advanced Magnetic Materials*, ed. by H. Kronmüller, S. Parkin, vol. 3 (Wiley, Chichester, 2007), pp. 1589–1613
76. E. Beaurepaire et al., Appl. Phys. Lett. **84**, 3465 (2004)
77. B. Koopmans et al., Phys. Rev. Lett. **95**, 267207 (2005)
78. G.M. Müller et al., Nat. Mater. **8**, 56 (2009)
79. U. Atxitia et al., Appl. Phys. Lett. **91**, 232507 (2007)

80. N. Kazantseva et al., Euro. Phys. Lett. **81**, 27004 (2008)
81. B. Koopmans et al., Nat. Mater. **9**, 259 (2010)
82. U. Atxitia, O. Chubykalo-Fesenko, Phys. Rev. B **84**, 144414 (2011)
83. A. Mekonnen et al., Phys. Rev. Lett. **107**, 117202 (2011)
84. S. Mizukami et al., Phys. Rev. Lett. **106**, 117201 (2011)
85. F.D. Longa et al., Phys. Rev. B **75**, 224431 (2007)
86. J.P. van der Ziel et al., Phys. Rev. Lett. **15**, 190 (1965)
87. A. Kirilyuk et al., Phil. Trans. R. Soc. A **369**, 3631 (2011)
88. A.V. Kimel et al., Nature **435**, 655 (2005)
89. K. Vahaplar et al., Phys. Rev. Lett. **103**, 117201 (2009)
90. T.A. Ostler et al., Nat. Commun. **3**, 666 (2012)

Chapter 6
Recent Topics for the Optical Properties in Liquid Crystals

Yoichi Takanishi

Abstract Liquid Crystals are equilibrium phases appearing between crystals and isotropic liquids, and have intermediate properties between them. Spontaneous meso-scale formation is also characteristic of liquid crystals, and comparatively easy fabrication of fine structures can be done. Because of these characters, liquid crystal is one of most successful functional organic materials from the viewpoint of application. That is obvious from the fact of the spread of flat panel display using liquid crystals. A liquid crystal flat panel displays utilize the light valve property of liquid crystals due to its optical anisotropy and easy tuning or alignment by external field such as electric field or mechanical stress. Such physical properties can be utilized for not only the displays but also various optical applications. In this chapter, I'd like to review recent topics of application using optical property of liquid crystals; photonic effects of cholesteric phase, new-type display mode using bent-core liquid crystals, and polymer stabilized blue phase and its application.

6.1 Introduction

Liquid Crystals (LCs) are equilibrium phases appearing between crystals and isotropic liquids, and have intermediate properties such as fluidity as liquids and anisotropy as crystals. Spontaneous meso-scale formation is also characteristic of LCs, and comparatively easy fabrication of fine structures can be done. Furthermore, LC is one of most successful functional organic materials from the viewpoint of application. That is obvious from the fact of the spread of flat panel display using LCs. A LC flat panel displays utilize the light valve property of LCs due to its optical anisotropy and easy tuning or alignment by an external field such as an electric field or a mechanical stress. Such physical properties can be utilized for not only the dis-

Y. Takanishi (✉)
Division of Physics and Astronomy, Department of Physics, Graduate School of Science, Kyoto University, Kitashirakawa-oiwakecho, Sakyo, Kyoto, 606-8502 Japan
e-mail: ytakanis@scphys.kyoto-u.ac.jp

Y. Aoyagi and K. Kajikawa (eds.), *Optical Properties of Advanced Materials*, Springer Series in Materials Science 168, DOI: 10.1007/978-3-642-33527-3_6,

plays but also various optical applications. In this chapter, I'd like to review recent topics of application using optical property of LCs. I hope this book will be useful for readers to make any new ideas for their future study plans.

6.2 Photonic Effects of CLC and Its Application

6.2.1 Cholesteric Liquid Crystal and Its Photonic Effects

Liquid crystal is an intermediate phase between an isotropic liquid and crystals, called a "mesophase" [1]. In the LC phase, molecules are spontaneously arranged with a certain degree of order which is intermediate between the perfect disorder of a liquid and long-rage order of a crystal. Liquid crystals are classified into many types by the degree of order and symmetry. In this chapter, treated liquid crystals are composed of rod-like molecules, and categorized into the thermotropic type, in which the mesophase are brought about the thermal process. The highest symmetric (most disordered) thermotropic liquid crystalline phase is the nematic phase. It has no long-range translational (positional) order like an isotropic liquid, but a relatively high degree of long-range orientational order of rod-like molecules (see Fig. 6.1a). This means that molecules spontaneously oriented as their long axes are almost parallel to each other. Hence macroscopically and uniformly aligned state is optically uniaxial, and strongly birefringent.

A variant of the nematic phase appears when whole or partial molecules are chiral and the mirror symmetry of the system is broken. In this system, neighboring molecules tend to align at a slight angle to one another, so that the twisting property is induced in the system. This phase is called cholesteric (or chiral nematic) phase. Molecular arrangement in the cholesteric phase are shown in Fig. 6.1b. Due to the chirality, twisted structure about an axis normal to the preferred directions of

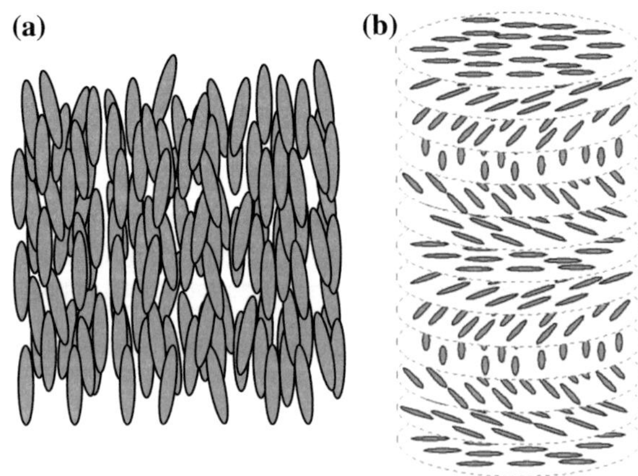

Fig. 6.1 Molecular arrangement of (**a**) nematic and (**b**) cholesteric phases

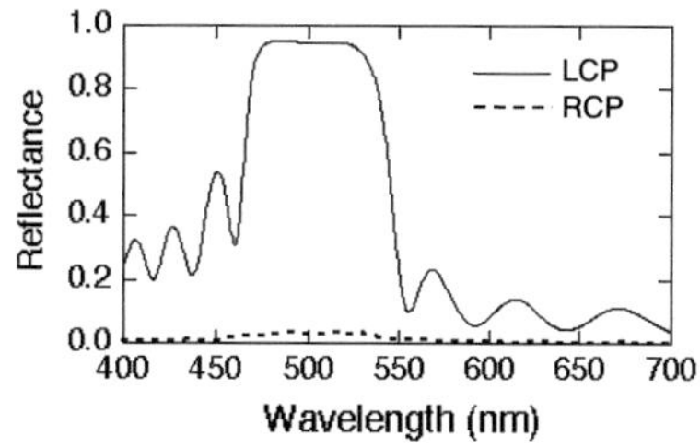

Fig. 6.2 Typical reflection spectra obtained from *left*-handed CLC when *left*- and *right*-handed circular polarized lights are incident

molecular long axes are formed, and the twist sense corresponds to the chiral sense of the system. Since twisted angle between neighboring molecules is slight and the energy of formation of twisted structures are lower than total energy of formation of the nematic orientation, it is relatively easy to form well-defined helical configuration if a small quantity of chiral molecules are contained in the system. The periodicity of helical structure is called helical pitch, which depends on the concentration of chiral molecules, temperatures, applied field strength and so on. Twist configuration of anisotropic molecules induces unique optical properties such as so-called selective reflection and optical rotation.

When the optical helical pitch is comparable to the visible light wavelength, specific light propagation occurs. That is called selective reflection; when light propagates through a CLC along its helical axis, one of the circularly polarized modes with the same handedness as the cholesteric helix and with the same wavelength as the helical pitch is totally reflected, while the other passes through without any loss. Although I neglect the details of mechanism of this property in this chapter, this phenomenon may be comprehensible if you can imagine that it is regarded as an analogue of X-ray Bragg reflection. The selective reflection condition is written as follows.

$$\lambda = np\cos\theta \tag{6.1}$$

Here λ is wavelength of the circular polarized light with the same sense of helical handedness, n average refractive index of the CLC, p physical helical pitch of the CLC and θ is an incident angle of light. Figure 6.2 shows the typical reflection spectra of CLC with left-handed helical sense. It is found that the left-handed circular polarized light (the same sense as the helical sense of CLC) is completely reflected, while right-handed circular polarized light (with opposite sense) is completely transparent. Moreover, it is noticeable that total reflection occurs not at one wavelength but within a certain range, $\Delta\lambda$,

$$\Delta\lambda = p \times \Delta n, \tag{6.2}$$

where Δn is optical birefringence ($= n_e - n_o$). It means that the photons of one of the circular-polarized lights with an energy range satisfying the above equation are prohibited to exist within CLC, and this optical property is regarded as a kind of photonic band gap [2]. The photonic band gap (PBG) was observed in crystals with periodic structures of optical wavelength, theoretically proposed by Yablonovitch [2]. Such crystals are called photonic crystals, and lights with a certain range of wavelength are prohibited. Since the origin of PBG is analogous to the origin of electron band gap of crystals, the $\omega - k$ dispersion relation of photon is efficiently same as the $E - k$ dispersion relation of electron.

6.2.2 Light Amplification Using CLCs

In general, at the edge of PBG, group velocity approaches to zero, so that the photonic density of state (DOS) increases. This is one of the most attractive optical properties of PBG because the increase in DOS is expected to apply to the light amplification. To apply this idea to CLCs, Kopp et al. prepared a CLC doped with an active medium (fluorescent dye), and first attained the laser action in CLCs [3]. Figure 6.3a shows the typical transmitted and emission spectra when the excitation intensity is low and efficiently high. When the excitation is low, broad fluorescent emission is observed, while a very sharp peak with narrow width less than 1 nm appears at just the lower-energy band edge when the excitation becomes high. Figure 6.3b shows the emission intensity as a function of excitation intensity, and we can find that emission intensity drastically increases at a certain excitation intensity, i.e., it shows the clear threshold property. Since there exists a laser active medium in the CLC structure, this laser action is of distributed feedback (DFB) type [4].

Since publication of the lasing action using dye-doped CLCs by Kopp et al., a lot of researchers have been involved in this field from various viewpoints. One of the

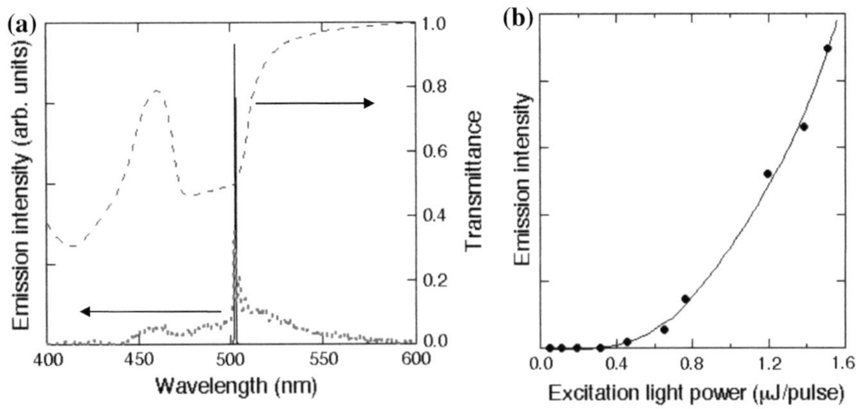

Fig. 6.3 **a** Typical transmitted (*broken line*) and emission spectra when the excitation intensity is low (*dotted line*) and efficiently high (*solid line*). **b** Emission intensity as a function of excitation light intensity

most important and urgent demands is how the lasing threshold can be lowered. Among various attempts, one of the solutions is the utilization of defect modes [5]. As the appearance of the defect level of electron in imperfect crystals, defect mode energy level appears by introducing the defect into the photonic structure. In this condition, an additional resonant mode appears inside the photonic band gap, and the localization of the light based on the defect mode occurs at the position of the defect. As a result, it is possible to make more effective optical confinement at a narrow space, and low threshold properties have been expected. The efficiency of the fabricated structure has been also theoretically studied and calculated. Details in the theoretical works are referred in Ref. [6, 7].

There are several methods for the introduction of the defect layer into the helical structure of the CLCs. Mainly two ways to introduce a defect have been proposed for CLCs, either by replacing a thin layer of the CLC by an isotropic material [8], or by introduction of a phase jump in the cholesteric helix at the intermediate region [9, 10]. Schimdtke et al. fabricated the phase jump in the cholesteric helix as a defect [9]; two layers of a highly cross-linked CLC polymer film are used to create a phase jump of 90° in the cholesteric structure. They certainly observed defect-mode lasing peak in the spectrum as shown in Fig. 6.4c, and confirmed the lower threshold (2.5 nJ) than that observed in the normal band-edge mode (50 nJ). At almost the same moment, Ozaki et al. also reported the very similar results [10]. Moreover, they numerically analyzed results of the tuning of the defect mode in CLCs, and actually attained the tuning of the defect mode by fabricating optically induced local deformation of the helix in the middle of the CLC doped with azobenzene using UV light [11].

On the other hand, the author's group also studied the defect mode laser action in CLCs by another technique. At first a phase-retardation defect-mode lasing was demonstrated by the construction of a cell composed of a dye-doped nematic defect layer sandwiched between polymer CLC (PCLC) films [12]. Moreover, by changing the dye-doped nematic layer to dye-doped low molecular CLC whose helical sense is opposite to that of PCLC films [13], the threshold becomes reduced by a factor of about 4 compared with that in a simple dye-doped low CLC cell (see Fig. 6.5). This result was supported by calculated results by 4×4 matrix method; an extremely high DOS peak is obtained at a wavelength where one of the defect modes and the band edge mode coincide with each other in the three-layered structure. Following this study, we studied the lasing property in another three-layered structure with the same PCLC helical handedness to that of a low-molecular weight CLC interlayer, and similar low threshold properties are also observed [14].

Later, the difference of lasing characteristics between two 3-layered cells was discussed [15]. Strictly speaking, threshold in the former structure (RLR) [12] is clearly lower than the latter (LLL) [14], and the far-field patterns of the lasing emission from the former cell shows a clear diffraction fringe pattern, which proves that the emission light has higher coherence (see Fig. 6.6a). In the latter cell, one of circularly polarized laser light is generated in the middle low CLC active layer. The lasing wavelength is within the band gap of outer PCLC films, and moreover the PCLC is not thick enough to act as leaky cavity mirrors for the low CLC. This is why the

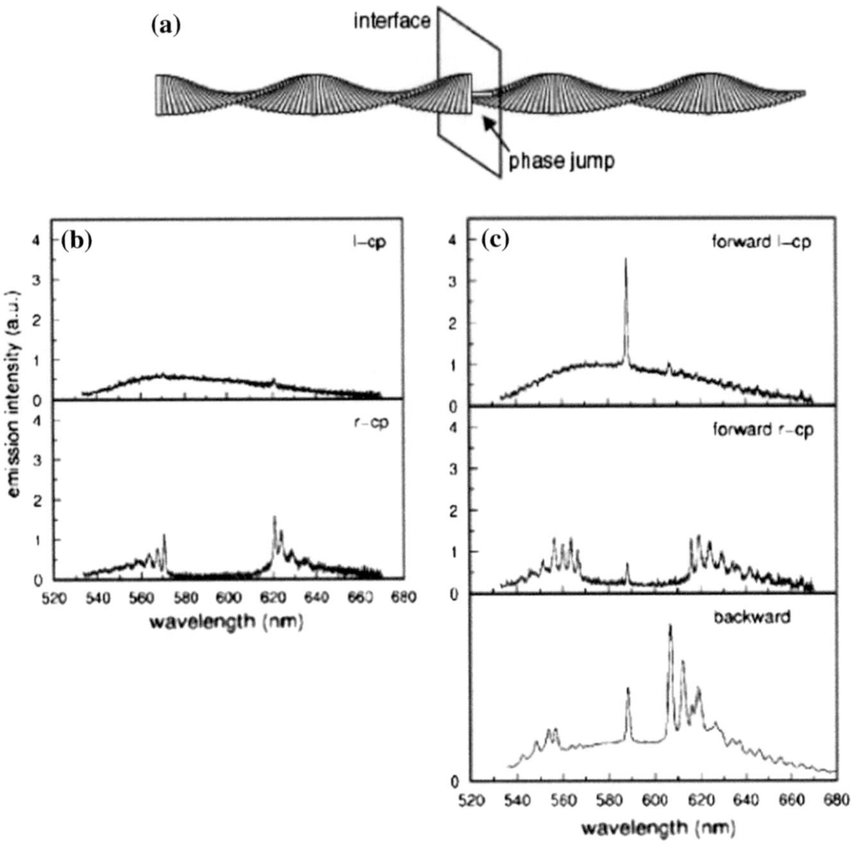

Fig. 6.4 **a** Image of the phase jump type defect structure of the CLCs, and emission spectra of cholesteric films; **b** single layer: l-cp (*top*) and r-cp (*bottom*) emission, and **c** double layer with phase jump of the pitch: l-cp (*top*) and r-cp (*middle*) forward emission, and total backward emission [9]

lower threshold is achieved compared with that in a normal cell. As for the former cell, the low-CLC acts as a defect layer for PCLC films. Actually, many defect states are seen within the whole band gap of PCLC in the DOS spectrum, while no defect states exist in the latter cell in this range. Together with the experimental fact that lasing occurs at wavelengths corresponding to the defect state within the band gap of the low-CLC, it can be concluded that the lasing from the former cell is due to a defect mode. This is the essential difference between them and explains the different threshold and cavity behavior.

Moreover, we fabricated multilayered CLC structures and evaluated the laser emission spectra as shown in Fig. 6.7 [15]. In their structure, multimode lasing spectra were observed, and calculated DOS results as a function of light wavelength are qualitatively consistent with the corresponding experimental spectra. From the result of emission intensity as a function of excitation intensity in each structure, it

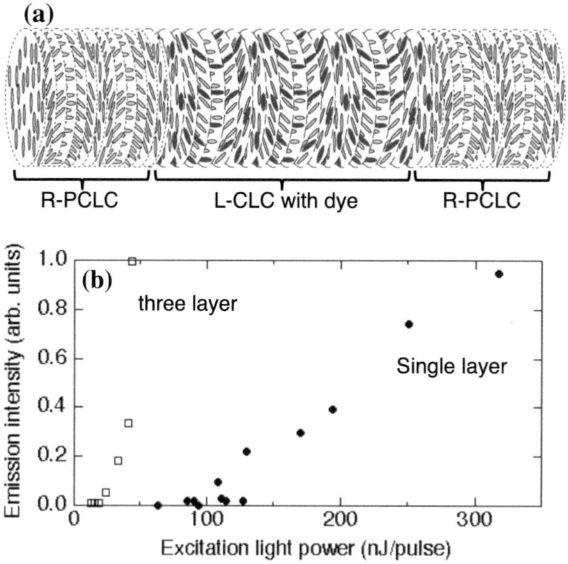

Fig. 6.5 **a** Schematic figure of CLC cell structure with polymeric dye-doped L-CLC sandwiched by two R-PCLC layers. **b** Threshold behaviors in the single layer and three layered cells [15]

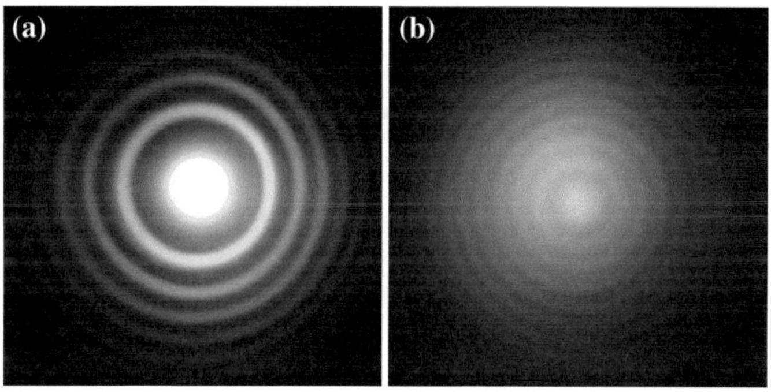

Fig. 6.6 Lasing pattern obtained from two 3-layered structures. **a** Intermediated layer has the same helical sense, and **b** intermediated layer has opposite sense [15]

was found that the lasing from five-layered structure was observed with an extremely reduced threshold by a factor of 10 compared with that in a simple dye-doped low-molecular-weight CLC cell.

Since the CLCs are soft materials as known well, it is easy to change the physical property by temperature, light, external fields such as an electric field and mechanical strain. Hence the tunability of CLC lasing is easily controlled. For example, Ozaki et al. presented the tunable laser action by applying an electric field [16]. Their used materials is a chiral SmC* compound, not CLCs. The chiral SmC* phase also has a helical structure whose pitch is of the same order as the wavelength of visible light

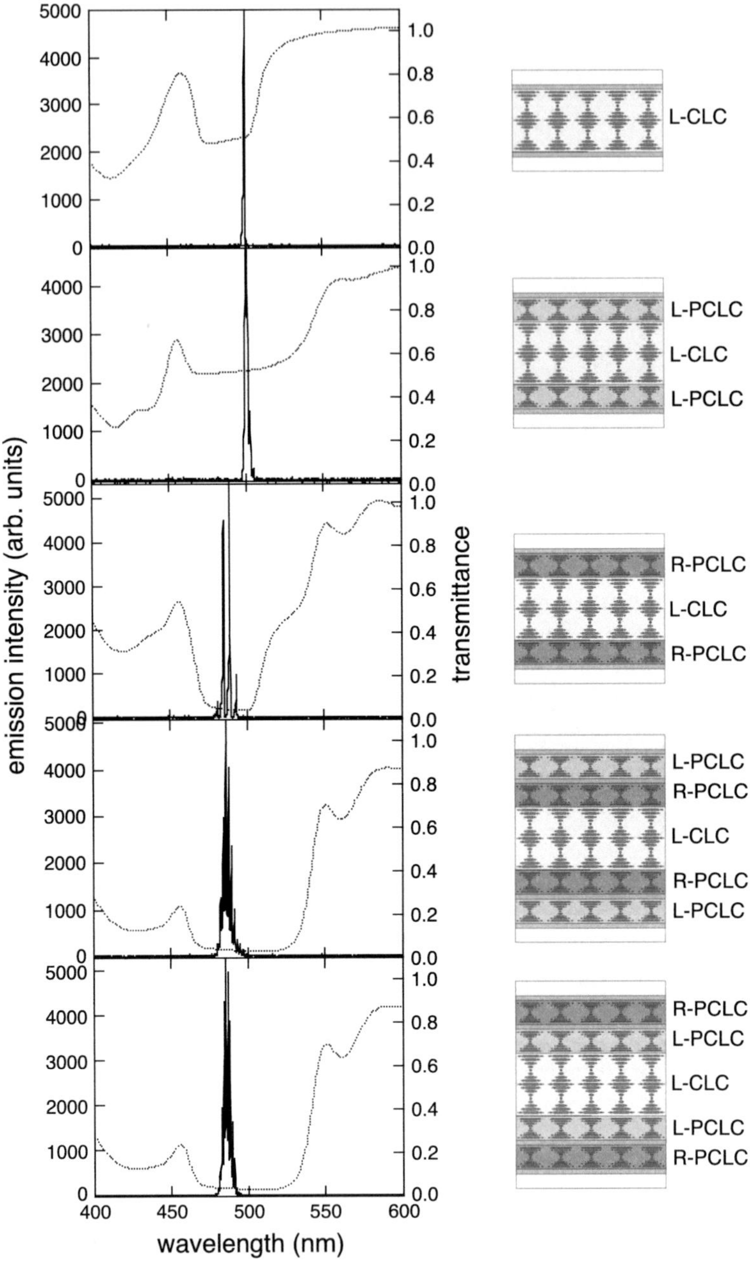

Fig. 6.7 *Left side* Transmittance and laser emission spectra obtained from multilayered CLC structures [15]. *Right side* Schematic figures of multilayered CLC structures

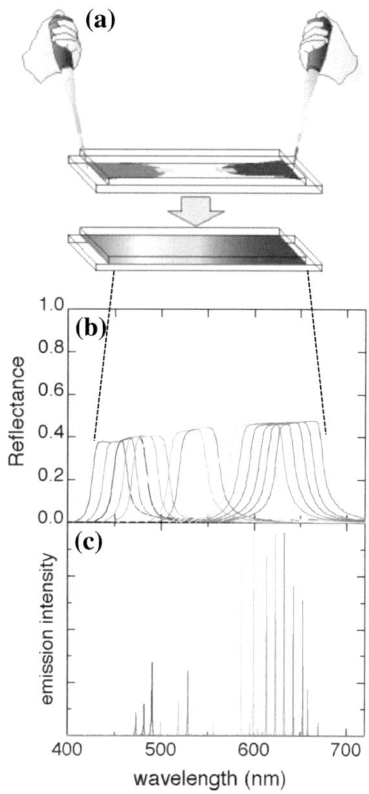

Fig. 6.8 Schematic figure of cholesteric-liquid-crystal cell with spatial gradient structures (**a**) and position-dependent reflection (**b**) and lasing spectra (**c**) of the dye-doped CLCs [21]

and exhibits selective reflection in the same manner as the CLCs. Chanishvili et al. [17, 18] and Huang et al. [19, 20] fabricated CLC structures with a spatial helical-pitch gradient and demonstrated lasing over a wide wavelength range. Unfortunately, there was a window of about 35 nm, where lasing was not possible. Moreover, six or more kinds of dyes were necessary for wide-band lasing. We also reported a CLC-DFB laser with spatial gradient structures not only of the helical pitch but also of the concentrations of two dyes. In our system, lasing was achieved covering from 470 to 670 nm without any windows by translating the cell with respect to a pumping beam without changing the excitation light wavelength because of the use of Förster energy transfer, as shown in Fig. 6.8 [21, 22]. In our proto-type system, spatial pitch gradient structure gradually relaxes to in a uniform one due to the molecular diffusion. Hence we changes CLC molecules to photo-curable ones, and succeeded to fix the pitch gradient structure [23].

Moreover, Ha et al. designed and fabricated a CLC system consisting of thin isotropic films and of polymeric CLC films, and demonstrated experimentally simultaneous red, green and blue reflections (multiple PBGs) using the single-pitched polymeric CLC films [24]. Figure 6.9a shows the schematic figure of multi-CLC

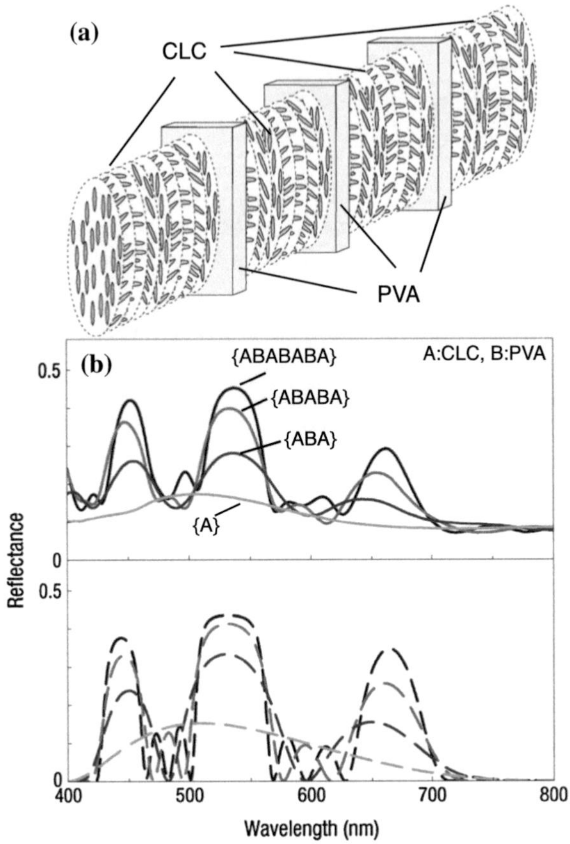

Fig. 6.9 **a** Schematic figure of multi-CLC system. **b** Experimental (*solid line*) and simulated (*dashed line*) reflection spectra for multilayered systems. A and B means PCLC and PVA layers, respectively [24]

system. By inserting PVA films as an isotrpic layers, it becomes possible to assemble this system. In advance, they showed by numerical simulation that simultaneous multi-colour reflection is possible by introducing fibonaccian-type phase defects, and obtained experimental reflection spectra are well simulated by calculations, as shown in Fig. 6.9b [24].

6.2.3 6-2-3 Optical Diode Using CLCs

Most of the researches of the photonic effect of CLCs are regarded with light amplification. Considering the optical property of CLCs, however, other unique application can be expected. One of examples is an optical rectification proposed by Hwang et al.

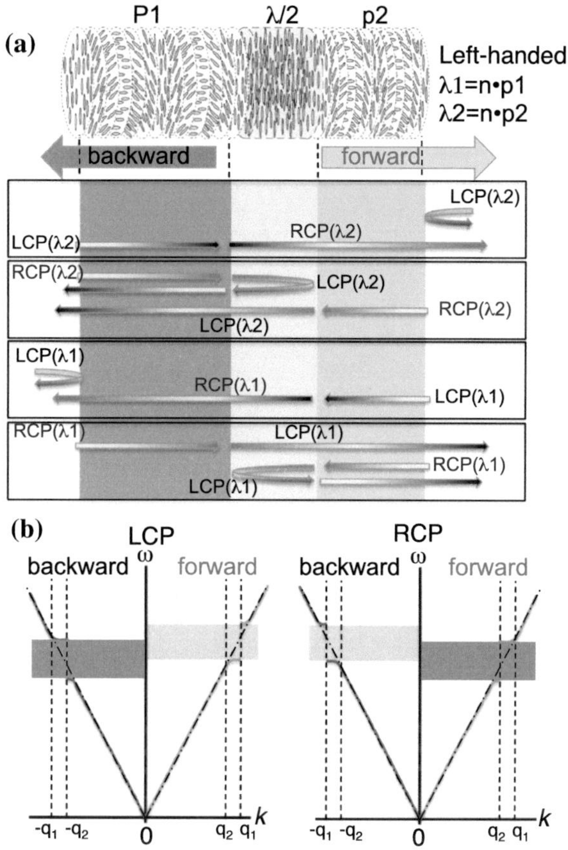

Fig. 6.10 **a** Schematic figure of optical diode proposed by Hwang et al., and the propagating relation for *right*-handed (RCP) and *left*-handed (LCP) circular polarize lights with the wavelength of λ_1 and λ_2. **b** Optical dispersion relation of this system [25]. In this case, CLC has a *left*-handed helix and $p_1 < p_2$

[25]. As you know, an diode is an electronic circuit, which permits a current in the forward direction and forbids it in the backward direction. They called their system an optical diode, because it can transmit light in one direction and forbid it in the reverse direction (non-reciprocal transmission).

The system forms a new type of PBG structure consisting of an anisotropic nematic layer sandwiched between two cholesteric liquid-crystal layers with different helical pitches. Let me consider a multilayer structure shown in Fig. 6.10a. It is called optical hetero-junction anisotropic structure (OHAS) in Ref. [25]. Outside CLC layers have the same handedness but different helical pitches, and an intermediate nematic layer works as a half wave phase retarder plate. The optical dispersion relation of this OHAS are shown in Fig. 6.10b. Lines drawn at positive and negative k regions correspond to dispersion for forward and backward propagation, respectively. PBG

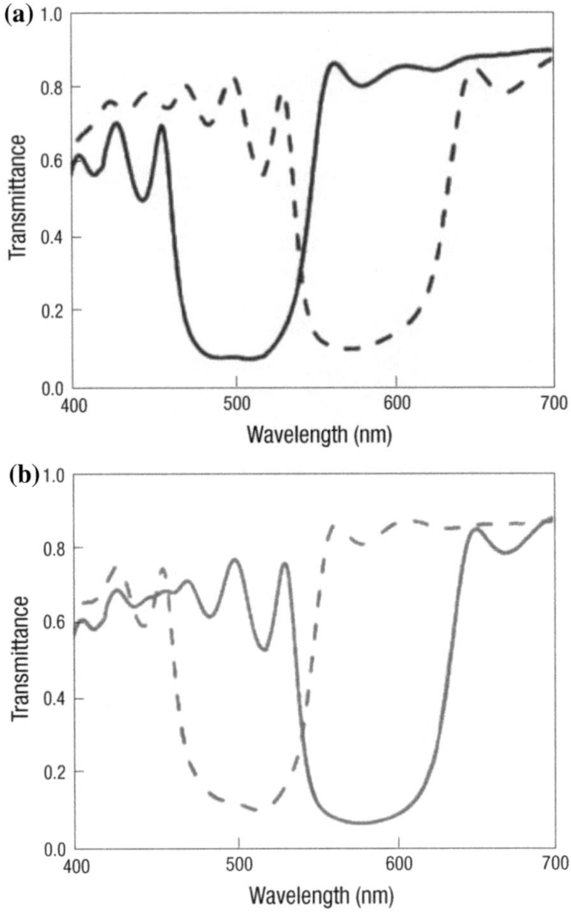

Fig. 6.11 Experimental transmission spectra of OHAS; (**a**) forward and (**b**) backward propagations respectively. Circularly polarized light with handedness the same as and opposite to that of CLC helix is indicated by *solid* and *broken curves*, respectively [25]

frequency range depends on the propagation direction. In case of Fig. 6.10b, LCP with dark gray (green) area frequency can propagate backward but not forward, while LCP with bright gray (orange) area frequency can propagate forward but not backward, which means optical diode behavior (non-reciprocal transmission) with respect to the wavelength. For two circular polarizations with opposite sense, the PBGs for forward and backward light propagations are interchanged. This relation originates from the exchange in handedness of the incident circular polarization on passing through an intermediate nematic layer as the half-wave phase retarder.

Figure 6.11 shows the transmission spectra measured for the fabricated OHAS cell in both forward (a) and backward (b) propagation directions and for a circularly polarized light with handedness the same as (solid curve) and opposite to (broken curve) that of CLC. With RCP and LCP lights, a reversed optical diode effect for forward and backward propagations is observed. This result is also supported by simulated results.

6.3 New Type Display Mode Using Bent-Core Liquid Crystals

As you know, liquid crystal displays (LCDs) have been developed day by day. LCD televisions (TVs) of large-size to reach 80 inches and of high qualities such as 3D mode etc are now commercially available. Development of an in-plane switching (IPS) mode [26] and a vertical alignment (VA) mode [27] instead of twisted nematic (TN) mode [28] can attain good performance such as high contrast ratio, wide viewing angle, and precise continuous gray levels. However, none of them have perfect performance in all problems, and engineers are still seeking for higher performance. One of serious problems would be fast response. Most of the displays use nematic liquid crystals, and it seems to be difficult to improve response time rapidly because of nematic viscosity and large magnitude of displacement of the molecular long axis under switching. Thus ferroelectric LC (FLC) [29, 30] or antiferroelectric LC (AFLC) [31, 32] modes, which exhibit fast response as much as microsecond order, had been developed since twenty years ago. In the conventional FLC and AFLC displays, and V-shaped switching (VS) mode [33], the molecules aligned parallel and the layer in the SmA phase forms almost normal to the substrates (Planar alignment). Tilted smectic phases such as ferroelectric SmC* and antiferroelectric SmCA* have temperature dependence of the layer spacing, a chevron layer structure along the substrate normal (vertical chevron) is formed due to the layer shrinkage from in the planar cell [34], and zig-zag defect and focal conics is frequently induced. Moreover the layer formation switches between chevron to bookshelf type during applying the electric field, and stripe domain texture caused by the formation of the chevron structure parallel to the substrate (horizontal chevron) [35] is observed. Because of the appearance of such defects decreases its contrast and destabilized the optical properties.

Recently Shimbo et al. proposed and demonstrated a totally new LCD mode using a recently discovered smectic A-like phase of banana-shaped liquid crystal molecules [36]. Bent-core (banana-shaped) mesogens, whose typical chemical structures are shown in Fig. 6.12, have opened a new field of LC science, and extensive researches

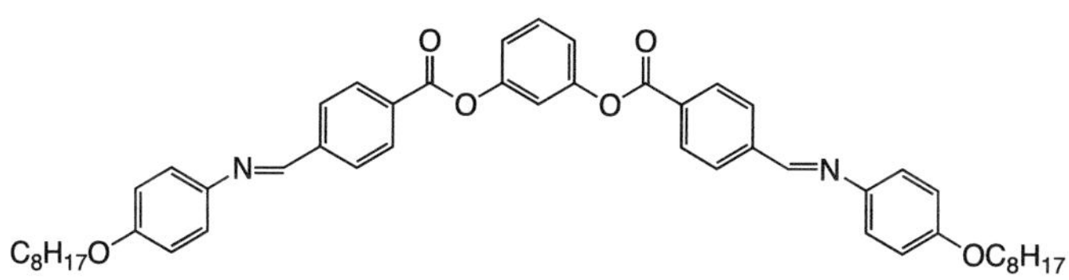

Fig. 6.12 Typical chemical structure of bent-core liquid crystal compounds

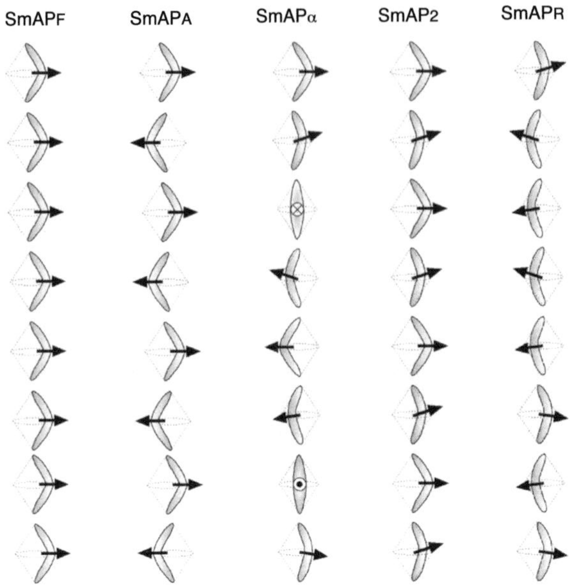

Fig. 6.13 Structures of polar SmA type phases of bent-core liquid crystals proposed in Ref. [40]. $SmAP_F$ and $SmAP_A$ has ferroelectiric and antiferroelectric order, respectively, and SmAPα resembles SmCα^*. SmAP2 has a bilayer structure, and the sign of the azimuthal phase difference between polarizations in neighboring layers alternates. In SmAPR, only interactions between neighboring layers are present, and the magnitude of the azimuthal phase difference is constant but its sign varies randomly from one pair of neighboring layers to another [40]

have been performed mostly from basic science viewpoints [37–39]. At the beginning of this study, it was found that the molecules exhibit specific smectic phases different from the rod-like smectic LCs, called $B1 \sim B8$. When the research proceeds and many bent-core molecules are synthesized later, and new nontilted phases have been also found in addition to the classic banana specific phases [40]. They tried to use one of these new phases as device of a new LC mode. These phases were discovered by Pociecha et al. [40], and it is apparently close to the SmA phase; i.e., optically uniaxial smectic phase without an electric field, but responds to an applied electric field like a ferroelectric phase. They at first called this mesophase SmX, and proposed three structures as SmAPR, SmAP2 and SmAPα, as shown in Fig. 6.13.

Principle of their new mode is depicted in Fig. 6.14. In this mode, homeotropic alignment is fabricated so that the smectic layer is parallel to the substrate. Figure 6.14b shows top and side views of the homeotropic cell, respectively, exhibiting the molecular and corresponding dipolar orientations during applying an electric field. The direction of the dipole of individual molecule is parallel to the molecular bent plane because of the molecular shape. This homeotropic cell is placed between crossed polarizers, and the applied electric field direction is parallel to the substrates and tilted by 45° to the polarizers. In the absence of an electric field, the phase is optically uniaxial due to the random orientation of molecular bent direc-

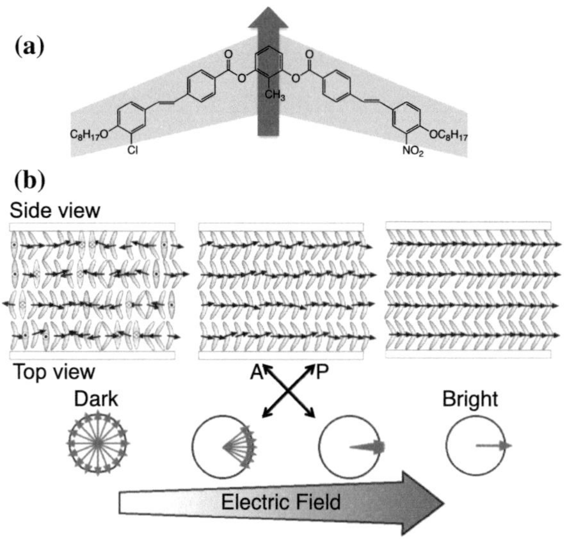

Fig. 6.14 Schematic images of novel mode using the SmA phase of bent-core liquid crystals [36]; (**a**) chemical structure and (**b**) molecular and dipolar arrangement under the application of in-plane electric field

tions, and the homeotropically aligned cells show a complete dark view under the optical geometry mentioned above because its optical axis is normal to the substrates. Hence the dipoles are randomly distributed, as shown in Fig. 6.14b (left). When an in-plane electric field is applied, the dipoles gradually orient toward the field direction. Because of the optical biaxiality of the bent-core molecules, such an orientational change gradually induces optical birefringence in the substrate plane. As a result, dark-and-bright display can be attained under the optical geometry. Since the molecular rearrangement gradually occurs, gray scale performance can be done.

When the electric field is applied to a randomly oriented dipole media, dipoles oriented to the field according to the Langevin process, but the quite high electric field should be necessary to align the dipoles of molecules because the dipole moment of individual molecule is small. Hence it is considered that this switching mode originates from the cooperative motion of bent molecules with relatively long-range orientational order of dipoles, based on a two-dimensional Langevin process. By their second harmonic generation measurement evaluating the nonlinear dipolar susceptibility, it is estimated that about 150 molecules will collectively response with respect to the applied field. In their report, the performances of this switching mode are as follows.

(1) Contrast ratio: 3000:1
(2) Response time: rise 130–240 μs depending on the field strength, fall 40 μs independent of the field strength.
(3) Continuous gray scale with a small threshold

(4) Wide viewing angle like IPS
(5) Relatively high field is still needed compared to nematic LCs.

These properties seem to be very attractive. On the other hand, there are some problems in the following.

(1) Lowering an electric field necessary to increase the brightness.
(2) Increase in optical birefringence.
(3) Widening the temperature range until room temperature.

To obtain high display performance, extensive efforts will be continuously made in particular for synthesizing new materials.

6.4 Polymer Stabilized Blue Phase and Its Application

6.4.1 Cholesteric Blue Phase

If the system has chirality, some specific structure and property appears. For example, when nematic liquid crystals have chirality by introducing chiral molecules, the twisting power occurs, and they form the one-dimensional helical structure whose axis is normal to the local director. It is called the cholesteric phase, mentioned in 6.2.1. In some chiral LCs systems, different mesophases are observed at a narrow temperature range between isotropic liquid and cholesteric phases [41]. Since at first this phase exhibited blue colored texture, it is called a blue phase. Since this color is caused by the Bragg reflection of light of its periodic structures together with helix like the cholesteric phase, other colored texture is also observed in the blue phase, depending on the compounds.

Since Rod-like molecules have cylindrical symmetry, twist axis can be locally extended radially in the plane normal to the long axis of a certain molecule when the chirality is strong, as shown in Fig. 6.15a. Such a structure is called double twist, and moreover, it forms a cylinder as a higher-order structure due to the director orientational order, called double twist cylinder. It is impossible to fit such double twist cylinders into three-dimensional space completely, so that line defects (disclination lines) are requested to relieve the elastic energy. Hence the macroscopic blue phase must be always accompanied with defects, so that it unstably appears over a narrow temperature range $\sim$1 K so far.

At present, three cholesteric blue phases are observed, assigned as BPI, BPII and BPIII in order with a low temperature. BPI and BPII have body-centered cubic (bcc) and face-centered cubic (fcc) symmetries, respectively, and BPIII possesses the same symmetry as the isotropic liquid phase [42].

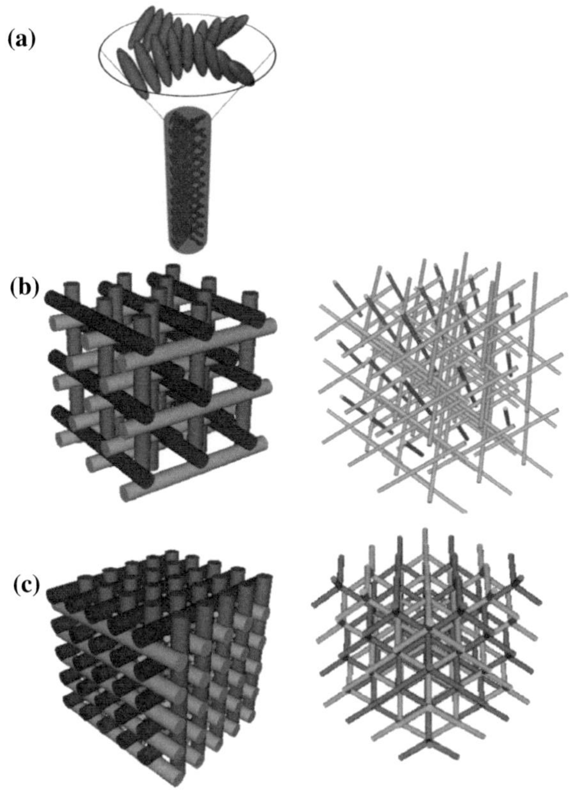

Fig. 6.15 Structure of (**a**) double twist cylinder, cholesteric *blue* phases (**b**) BPI and (**c**) BPII. *Right* side figures of (**b**) and (**c**) shows the corresponding arrangement of disclination line defects. These figures are drawn by Mr. Fumitaka Sakai

6.4.2 Attempt of Widening Temperature of the Blue Phase: Polymer Stabilized Blue Phase

From the viewpoint of practical applications, blue phases possess great potential because of their unique characteristics, such as electrically controllable Bragg diffraction of visible light [43–46] and a photonic band. As mentioned the former section, however, three-dimensional structure of the blue phase is formed by a very delicate balance between the chiral intermolecular interaction and the topological defect requirement, resulting in a frustrated system. Hence the limitation of the available temperature range—only a few kelvin—of blue phases has always been a problem. In this section, I will introduce very sophisticated techniques to widen temperature of the blue phase.

Considering that the structure of the blue phases is stabilized by its coexistence with disclination lines, Kikuchi et al. tried to stabilize the disclination lines [47].

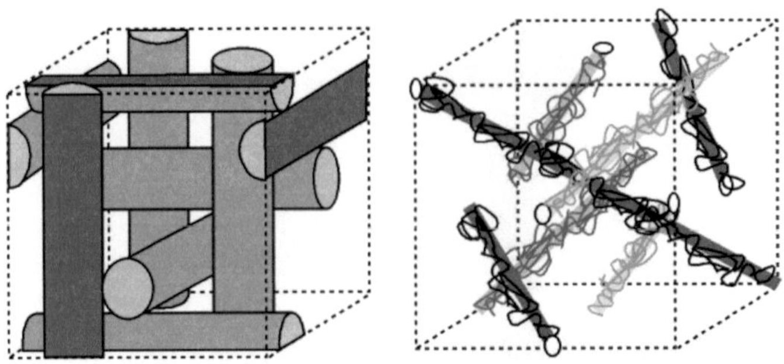

Fig. 6.16 Structure of polymer stabilized *blue* phase I [47]; arrangement of double twist cylinders (**a**) and aggregation state of photo-polymer network around the *disclination lines* (**b**)

They mixed acrylate monomers into liquid crystalline mixtures showing the blue phase in a certain ratio, and UV light was carefully irradiated in the blue phase for the photo-polymerization during the polarized microscope observation of texture not to induce the phase transition to the cholesteric phase. Photo-curable polymers are regarded as an impurity for the liquid crystal molecules, so they are localized around the disclination line (Fig. 6.16). In this way, they successfully expanded the temperature range of the blue phase up to more than 60 K.

The free energy per unit length of a single disclination line F for a lattice of defects is expressed as follows [48, 49],

$$F = K \ln\left(\frac{R_0}{R_c}\right) + \frac{1}{2}\int K\nabla \cdot [(\boldsymbol{n}\cdot\nabla)\boldsymbol{n} - \boldsymbol{n}(\nabla\cdot\boldsymbol{n})]d\boldsymbol{r}^3 + \alpha(T_{\mathrm{iso}} - T)\pi R_0^2 + 2\pi\sigma R_c \tag{6.3}$$

The first term is the elastic energy with respect to the defect, and K, R_c, and R_0 are Frank elastic constant, a disclination core radius and a cutoff radius size, respectively. The second term is an additional elastic energy term, which is specific for the blue phase. Here $\boldsymbol{n}$ is director. The third term is the energy for the defect core, where T_{iso} is the phase transition temperature to the isotropic state. Since the order at the disclination core is isotropic even in the blue phase, the energy should be lost when the disclination exists. The last term is the energy of the interface between the defect core and the chiral molecules outside the defect core. Here σ is the surface tension.

The structure of the normal blue phase, is stabilized by its coexistence with disclination lines as mentioned above, and the cores of disclination lines are assumed to be isotropic liquid state of mesogenic molecules. In general, a polymer is more favorite to isotropic order field, so that it is more miscible with an isotropic phase than a liquid crystal phase with orientational order. Hence photo-polymerized medium is considered to localize very close to the disclination cores as shown in (Fig. 6.16b). In such a situation, photo-polymer works as a template of the cubic lattice formed in the blue phase [47].

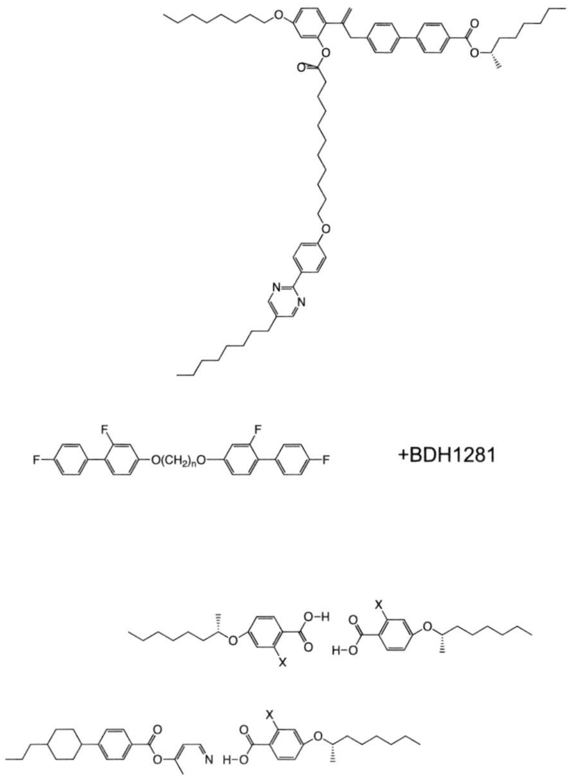

Fig. 6.17 Chemical structure of molecules exhibiting *blue* phase with a wide temperature [52–54]

The most stable steric conformation of flexible polymer chains dissolved in a liquid is a random coil. Therefore, the polymer chains localized in the disclination core should be in a random coil conformation, which contributes to disturbing the orientational order of the liquid crystal molecules. The blue phase of the composite therefore does not require thermal energy to keep the disclination core isotropic at a temperature below the phase transition temperature from the isotropic phase. This structure has been confirmed using various techniques such as X-ray, confocal miscroscope [50] and SEM [51].

After publishing this paper, several studies to widening temperature of the blue phase are reported. Yoshizawa et al. considered that the coupling between molecular biaxiality and chirality, and syntheseized T-shaped molecules (Fig. 6.17), and found that they shows the blue phase with a wide temperature range (ca. 20 K) [52]. The compounds showing the blue phase with wide temperature range reported by Coles et al. [53] and He et al. [54] have fluorine atoms at the lateral position of molecules, though fundamental chemical structures are different. Karatairi et al. found the expansion of BPIII temperature range by doping of surface-functionalized nanopar-

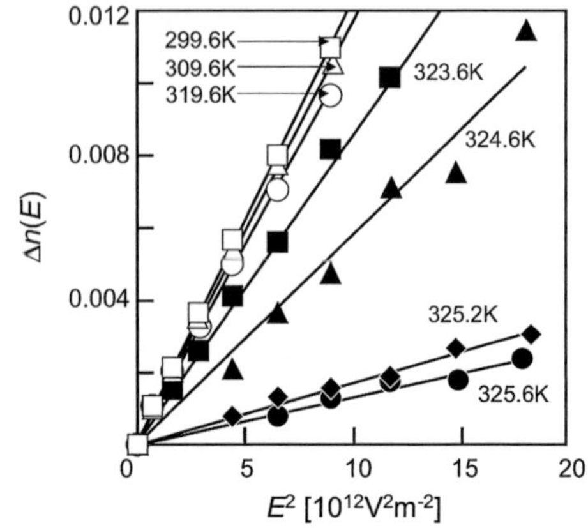

Fig. 6.18 Optical birefringence as a function of electric field of the polymer stabilized *blue* phase at various temperatures [57]

ticles [55]. Taushanoff et al., Koa et al.? reported the appearance of BPIII with a relatively low and wide (over 20 K) temperature range in nematic achiral bent-core liquid crystals doped with a high twisting power chiral material [56]. Since theoretical researches also increases since the development of polymer stabilized blue phase, this field will advance in the near future.

6.4.3 Electro-Optical Property of Polymer Stabilized Blue Phase and Its Application for Display

As mentioned above, the blue phase has cubic symmetry, so that optical property shows an isotropic order. Hence if the blue phase has the cubic periodicity whose lattice constant is less than visible-light wavelength, it should show completely dark under the crossed polarizers. Haseba et al. studied the electro-optic property in the polymer-stabilized blue phase [57]. Optical birefringence as a function of electric field was shown in Fig. 6.18. Δn is quadratically proportional to the applied field, which clearly responds to the optical Kerr effect. The Kerr effect, also called the quadratic electro-optic effect (QEO effect), is a change in the refractive index of a medium by applying an electric field E, and the magnitude of Δn is expressed as follows.

$$\Delta n = \lambda K E^2 \tag{6.4}$$

Here λ is the wavelength of light and K is the Kerr constant. In their results, K in the blue phase is $3.7 \times 10^{10}\,\mathrm{mV}^{-2}$, which is about 170 times as large as that of nitrobenzene, which is one of typical organic materials with a large Kerr coefficient (Fig. 6.18).

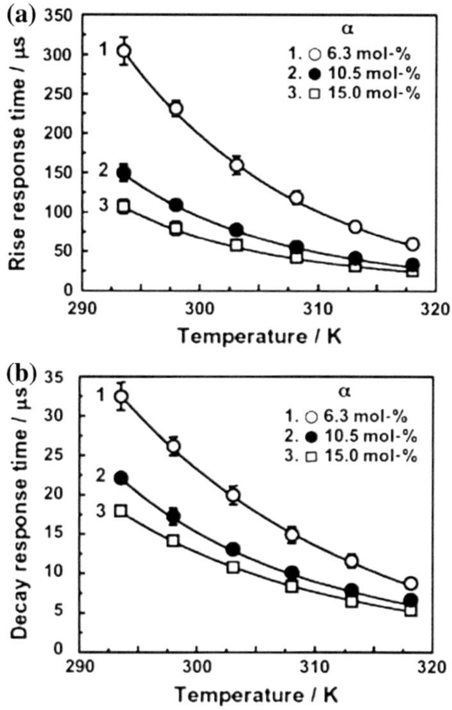

Fig. 6.19 Optical response time as a function of temperature; (**a**) rise time and (**b**) decay time [58]

What is the origin of the large optical Kerr effect in the blue phase? Cubic lattices of the blue phase are distorted when an electric field is applied, and this is known as an electrostriction. This process induces optical birefringence, but the shift of Bragg reflection due to the electrostriction is very small. In their results, the shift of (110) Bragg reflection is less than 1 % in the application of an electric field of 4.9×10^6 V/m, so that the distortion of lattice is not the main origin for the large Kerr effect. Therefore, a local molecular reorientation to unwind or deform the helix is considered to be mainly contributed to the induced optical birefringence [58].

Figure 6.19 shows optical response time as a function of temperature. The direction of the optical axis of the blue phase parallel to the electric field was placed at 45° to transparent easy axes's of crossed polarizers. Response time increases with decreasing temperature, but even at room temperature, it is the order of microseconds, much shorter than that in the nematic phase. It was also found that response time becomes faster with increasing the monomer concentration. Afterwards, it was also revealed that the effect of the chiral pitch on the response time was dependent on the type of electric field-induced birefringence of the blue phase, such as local director reorientation and electrostriction; the local director reorientation process shows very fast response at sub-ms while, in the case of the electrostriction process, the response time becomes a few ms [59]. Anyway, lowering an electric field necessary to increase the brightness will be future subjects.

References

1. For example, P.G. De Gennes, J. Prost, *The Physics of Liquid Crystals*, 2nd edn. (Clarendon, Oxford, 1993).
2. E. Yablonovitch, Phys. Rev. Lett. **58**, 2059 (1987)
3. V.I. Kopp, B. Fan, H.K. Vithana, A.Z. Genack, Opt. Lett. **23**, 1707 (1998)
4. H. Kogelnik, C.V. Shank, Appl. Phys. Lett. **18**, 152 (1971)
5. P.R. Villeneuve, S. Fan, J.D. Joannopoulos, Phys. Rev. B **54**, 7837 (1996)
6. For example, Y.-K. Ha, Y.-C. Yang, J.-E. Kim, H.Y. Park, C.-S. Kee, H. Lim, J.-C. Lee, Appl. Phys. Lett. **79**, 15 (2001)
7. J. Schmidtke, W. Stille, Eur. Phys. J. B **31**, 179 (2003)
8. Y.C. Yang, C.S. Kee, J.E. Kim, H.Y. Park, Phys. Rev. E: Stat. Phys., Plasmas, Fluids, Relat. Interdiscip. Top. **60**, 6852 (1999)
9. J. Schmidtke, W. Stillem, H. Finkelmann, Phys. Rev. Lett. **90**, 083902 (2003)
10. R. Ozaki et al., JJAP **42**, L472 (2003)
11. H. Yoshida, C.H. Lee, Y. Miura, A. Fujii, M. Ozaki, Appl. Phys. Lett. **90**, 071107 (2007)
12. M.H. Song, B. Park, K.-C. Shin, T. Ohta, Y. Tsunoda, H. Hoshi, Y. Takanishi, K. Ishikawa, J. Watanabe, S. Nishimura, T. Toyooka, Z. Shu, T.M. Swager, H. Takezoe, Adv. Mater. **16**, 779–783 (2004)
13. M.H. Song, N.Y. Ha, K. Amemiya, B. Park, Y. Takanishi, K. Ishikawa, J.W. Wu, S. Nishimura, T. Toyooka, H. Takezoe, Adv. Mater. **18**, 193 (2006)
14. Y. Takanishi, N. Tomoe, N.Y. Ha, T. Toyooka, S. Nishimura, K. Ishikawa, H. Takezoe, Jpn. J. Appl. Phys. **46**, 3510 (2007)
15. Y. Takanishi, Y. Ohtsuka, G. Suzaki, S. Nishimura, H. Takezoe, Opt. Express **18**, 12909 (2010)
16. M. Ozaki, M. Kasano, T. Kitasho, D. Ganzke, W. Haase, K. Yoshino, Adv. Mater. **15**, 974 (2003)
17. A. Chanishvili, G. Chilaya, G. Petriashvili, R. Barberi, R. Bartolino, G. Cipparrone, A. Mazzulla, R. Oriol, Adv. Mater. **16**, 791 (2004)
18. A. Chanishvili, G. Chilaya, G. Petriashvili, R. Barberi, R. Bartolino, G. Cipparrone, A. Mazzulla, R. Gimenez, L. Oriol, M. Pinol, Appl. Phys. Lett. **86**, 051107 (2005)
19. Y. Huang, Y. Zhou, S.-T. Wu, Appl. Phys. Lett. **88**, 011107 (2006)
20. Y. Huang, L.-P. Chen, C. Doyle, Y. Zhou, S.-T. Wu, Appl. Phys. Lett. **89**, 111106 (2006)
21. K. Sonoyama, Y. Takanishi, K. Ishikawa, H. Taekzoe, Jpn. J. Appl. Phys. Exp. Lett. **46**, L874 (2007)
22. K. Sonoyama, Y. Takanishi, K. Ishikawa, H. Takezoe, Appl. Exp. Lett. **1**, 032002 (2008)
23. T. Manabe, K. Sonoyama, Y. Takanishi, K. Ishikawa, H. Takezoe, J. Mater. Chem. **18**, 3040 (2008)
24. N.Y. Ha, Y. Ohtsuka, S.M. Jeong, S. Nishimura, G. Suzaki, Y. Takanishi, K. Ishikawa, H. Takezoe, Nat. Mater. **7**, 43 (2008)
25. J. Hwang, M.H. Song, B. Park, S. Nishimura, T. Toyooka, J.W. Wu, Y. Takanishi, K. Ishikawa, H. Takezoe, Nat. Mater. **4**, 383 (2005)
26. R.A. Soref, J. Appl. Phys. **45**, 5466 (1974)
27. M.F. Schiekel, K. Fahrenschon, Appl. Phys. Lett. **19**, 391 (1971)
28. M. Schadt, W. Helfrich, Appl. Phys. Lett. **18**, 127 (1971)
29. R.B. Meyer, L. Liebert, L. Strzelecki, P. Keller, J. Phys. Lett. **36**, 69 (1975)
30. N.A. Clark, S.T. Lagerwall, Appl. Phys. Lett. **36**, 899 (1980)
31. A.D.L. Chandani, E. Gorecka, Y. Ouchi, H. Takezoe, A. Fukuda, Jpn. J. Appl. Phys. **28**, L1265 (1989)
32. Y. Yamada, N. Yamamoto, K. Mori, K. Nakamura, T. Hagiwara, Y. Suzuki, I. Kawamura, H. Orihara, Y. Ishibashi, Jpn. J. Appl. Phys. **29**, 1757 (1990)
33. S. Inui, N. Iimura, T. Suzuki, H. Iwane, K. Miyachi, Y. Takanishi, A. Fukuda, J. Mater. Chem. **6**, 671–673 (1996)
34. T.P. Rieker, N.A. Clark, G.S. Smith, D.S. Parmar, E.B. Sirota, C.R. Safinya, Phys. Rev. Lett. **59**, 2658 (1987)

35. I. Dierking, F. Gieselmann, J. Schacht, P. Zugenmaier, Liq. Cryst. **19**, 179 (1995)
36. Y. Shimbo, Y. Takanishi, K. Ishikawa, E. Gorecka, D. Pociecha, J. Mieczkowski, K. Gomola, H. Takezoe, Jpn. J. Appl. Phys. (Express letter) **45**, L282 (2006)
37. T. Niori, T. Sekine, J. Watanabe, T. Furukawa, H. Takezoe, J. Mater. Chem. **6**, 1231 (1996)
38. D.R. Link, G. Natale, R. Shao, J.E. Maclennan, N.A. Clark, E. Korblova, D.M. Walba, Science **278**, 1924 (1997)
39. H. Takezoe, Y. Takanishi, Jpn. J. Appl. Phys. (Invited Review Paper) **45**, 597 (2006)
40. D. Pociecha, M. Cepic, E. Gorecka, J. Mieczkowski, Phys. Rev. Lett. **91**, 185501 (2003)
41. Q. Lehman, Z. Phys. Chem. **56**, 750 (1906)
42. For example, P.P. Crooker, Mol. Cryst. Liq. Cryst. **98**, 31 (1983)
43. V.E. Dmitrienko, Liq. Cryst. **5**, 847 (1989)
44. P.R. Gerber, Mol. Cryst. Liq. Cryst. **116**, 197 (1985)
45. H.G. Coles, H.F. Gleeson, Mol. Cryst. Liq. Cryst. **167**, 213 (1989)
46. G. Heppke, B. Jerome, H.-S. Kitzerow, P. Pieranski, J. Phys. (Paris) **50**, 2991 (1991)
47. H. Kikuchi, M. Yokota, Y. Hisakado, H. Yang, T. Kajiyama, Nat. Mater. **1**, 64 (2002)
48. J.P. Sethna, *Theory and Application of Liquid Crystals* ed by J.L. Ericksen, D. Kinderlehrer (Springer, New York, 1987) pp. 305
49. D.C. Wright, N.D. Mermin, Rev. Mod. Phys. **61**, 385 (1989)
50. K. Higashiguchi, K. Yasui, H. Kikuchi, J. Am. Chem. Soc. **130**, 6326 (2008)
51. T. Noma, M. Ojima, H. Asagi, Y. Kawahira, A. Fujii, M. Ozaki, H. Kikuchi, e-J. Surf. Sci. Nanotech. **6**, 17 (2008)
52. A. Yoshizawa, M. Sato, J. Rokunohe, J. Mater. Chem. **15**, 3285 (2005)
53. H.J. Coles, M.N. Pivnenko, Nature **436**, 997 (2005)
54. W. He, G. Pan, Z. Yang, D. Zhao, G. Niu, W. Huang, X. Yuan, J. Guo, H. Cao, H. Yang, Adv. Mater. **21**, 2050 (2009)
55. E. Karatairi, B. Rozic, Z. Kutnjak, V. Tzitzios, G. Nounesis, G. Cordoyiannism, J. Thoen, C. Glorieux, S. Kralj, Phys. Rev. E **81**, 041703 (2010)
56. S. Taushanoff, K.V. Le, J. Williams, R.J. Twieg, B.K. Sadashiva, H. Takezoe, A. Jakli, J. Mater. Chem. **20**, 5893 (2010)
57. Y. Haseba, H. Kikuchi, T. Nagamura, T. Kajiyama, Adv. Mater. **17**, 2311 (2005)
58. Y. Hisakado, H. Kikuchi, T. Nagamura, T. Kajiyama, Adv. Mater. **17**, 96 (2005)
59. H. Choi, H. Higuchi, H. Kikuchi, Soft Matter **7**, 4252 (2011)

Chapter 7
Materials for Organic Light Emitting Devices

Katsuhiko Fujita

Abstract Organic light emitting devices are self-emitting light source based on radiative decay of exciton generated by the carrier recombination on an organic molecule, and utilized for various applications. The device efficiency has been improved drastically in recent a couple of decades. This chapter deals with the mechanism and materials of the devices and the improvement in luminance efficiency.

7.1 General Remarks

Organic light emitting devices (OLED) are attracting much attention for light and thin flat panel display [1]. The typical device structure is illustrated schematically in Fig. 7.1. It has a simple multilayer structure where an organic semiconductor thin film is sandwiched with an anode and a cathode. When the voltage is applied between the electrodes, the organic film gives luminance. The electric current injected into the film leads the excited state in the organic molecules and the excess energy is discharged as luminance upon radiative decay back to the ground state. The luminance color is peculiar to the organic molecule. Though it is often called organic electroluminescence (EL), it is different from the inorganic EL where the hot electrons accelerated by the applied electric field supply the energy for the luminance. The organic EL is caused by the charge carrier injection from both electrodes into the organic film.

The organic EL phenomenon has been known since 1950s. It was observed that anthracene single crystal gave blue luminance when high voltage was applied [2]. It was already revealed that the positive and negative charge carriers were injected from the electrodes and that the recombination of the carriers in the organic molecule

K. Fujita
Institute for Materials Chemistry and Engineering, Kyushu University,
Kasuga, Fukuoka 816-8580 , Japan
e-mail: katsuf@asem.kyushu-u.ac.jp

Y. Aoyagi and K. Kajikawa (eds.), *Optical Properties of Advanced Materials*,
Springer Series in Materials Science 168, DOI: 10.1007/978-3-642-33527-3_7,

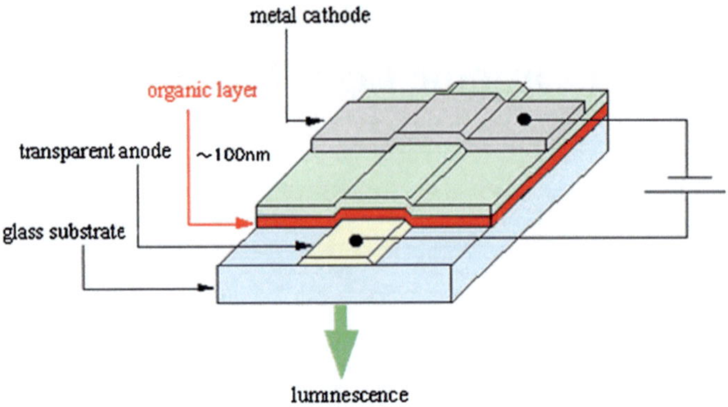

Fig. 7.1 Schematic illustration of typical OLED structure

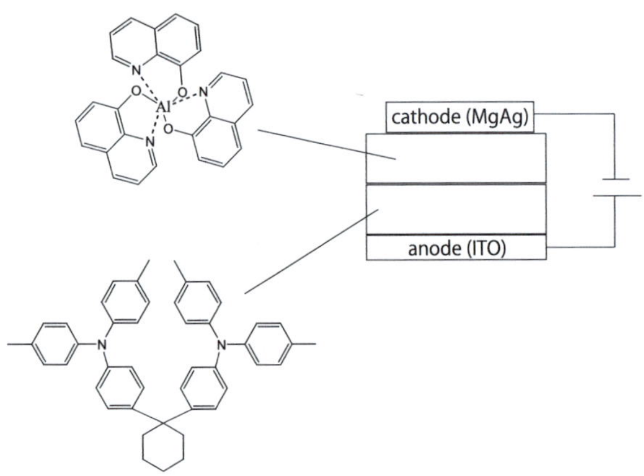

Fig. 7.2 Double layered OLED structure

can generate the excited state followed by the radiative decay. The EL spectrum was corresponding to the fluorescence spectrum of the anthracene. This basic principle of the device operation was already established at the very early study in 1960s. However, the high electric field was required since the relatively thick single crystals were used. An ultra-thin film of the anthracene prepared by the vacuum deposition was applied for the organic EL to reduce the driving voltage [3]. In 1987, it was reported that a layered organic film of a hole transport material, HTM, and an electron transport material, ETM, (Fig. 7.2) could give dramatic improvement in the luminance efficiency [4]. The observation was the substantial trigger for the development of the modern OLEDs.

7.2 Operation Principle

The energy diagram of the double layered OLED is shown in Fig. 7.3. The organic film with the total thickness of typically 100 nm is sandwiched by an anode with high work function and a cathode with low work function. Holes are injected from the anode to the hole transport level of the HTM and electrons are injected from the cathode to the electron transport level of the ETM. In general the hole transport level and the electron transport level are located very close to the highest occupied molecular orbital (HOMO) and the lowest unoccupied molecular orbital (LUMO) of an individual organic molecule, respectively. They are often used expediently as the transport levels. Organic thin films in an amorphous state should be molecular assembly based on very weak interaction through van der Waals force. It is hardly expected that the wide band is formed in the film. The carrier transport should be based on electron hopping among the transport levels of the individual molecules. The injected holes and electrons travel across the each transport materials and recombined on a molecule, resulting in excitons. The fraction which cannot meet the counter carrier goes through to the opposite end of the film and should be vanished.

The excitons generated by the carrier recombination can be divided in two spin states, singlet excitons and triplet excitons. These excitons undergo the numerous decay routes including radiative decay. Since the transition from the triplet excited state to the ground state is an inter system crossing and therefore, forbidden in ordinary molecules, the rate of the radiation decay from the triplet excitons, phosphorescence, is quite low. The prompt fluorescence by the radiative decay of the singlet excitons would contribute mainly to the EL in ordinary fluorescence OLEDs. However, the generation ratio of singlet exciton and triplet exciton would be 1:3 due to the spin statistics [5]. Only 25 % of excitons can contribute to the EL.

The photons have to come out of the device to contribute the EL. However, significant fraction of the photons generated by the radiative decay is confined in

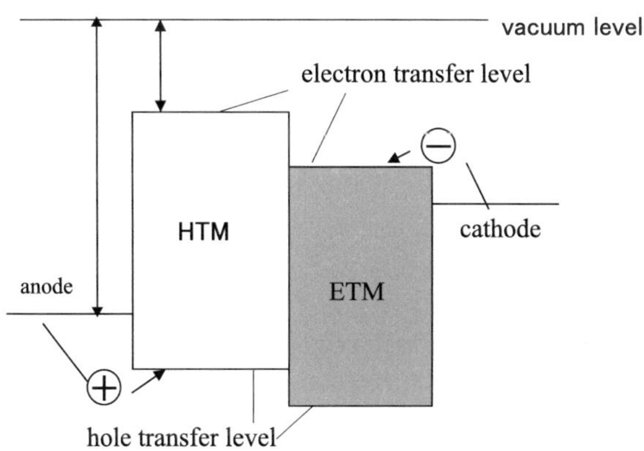

Fig. 7.3 Energy diagram of the double layered OLED

the device and does not come out due to the relatively high refractive index of the organic films.

The operation mechanism of OLEDs can be analyzed in the following elementary processes [5].

1. carrier injection from the anode and the cathode to HTM and ETM, respectively
2. carrier transportation based on electron hopping among transport levels
3. carrier recombination on a molecule
4. generation of singlet and triplet excitons
5. radiative decay of the excitons
6. photon extraction out of the device.

The external quantum efficiency (EQE) of EL is defined as the ratio between the photons coming out of the device and the electrons injected into the device, and can be expressed as the product of the probability at determining elementary process.

$$\mathrm{EQE} = \gamma\beta\phi_{\mathrm{f}}\chi \tag{7.1}$$

where γ: charge balance factor, β: probability of emissive species production, ϕ_{f}: quantum yield of luminescence, χ: is coupling-out factor dealing with the photon extraction (Fig. 7.4).

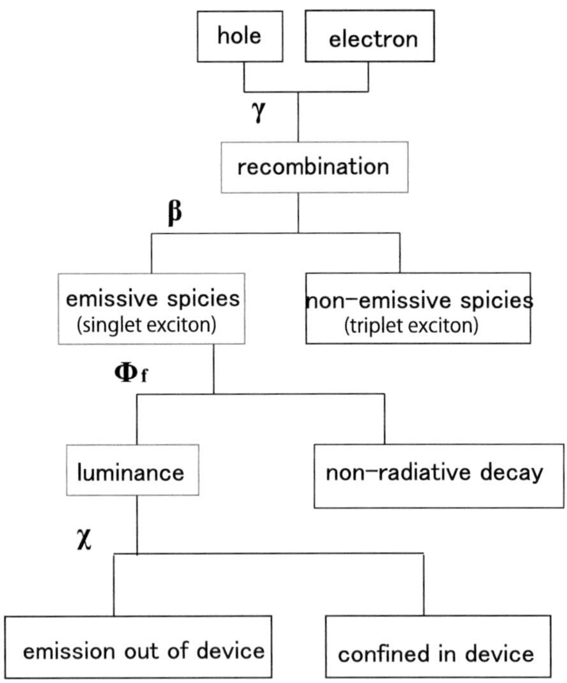

Fig. 7.4 Flow chart of elementary processes in OLED operation

The history of OLED development is a record of the struggles to improve each factor.

7.3 Charge Carrier Injection, Transportation and Recombination

7.3.1 *Carrier Injection*

Most of organic molecules have a closed shell structure. It is, therefore, not expected to form an interfacial state caused by the dangling bonds which are often seen in inorganic crystals or amorphous. The energy gap between carrier transport level and the work function of the electrode is basically maintained after the electrode attaches to the organic materials. The energy gaps can be as high as 0.7 eV in many cases. Even under high electric field over MVcm^{-1}, the carrier injection should be a difficult event. The two types of injection mechanisms to overcome the energy gap are commonly considered, Schottky thermal emission and Flower-Nordhiem tunneling [6] (Fig. 7.5). Schottky thermal emission

$$J = \frac{4\pi q m^* k}{h^3} T^2 \exp\left(-\frac{q\phi_\mathrm{B}}{K_\mathrm{B}T}\right)\left[\exp\left(\frac{qV}{K_\mathrm{B}T}\right) - 1\right] \tag{7.2}$$

Flower-Nordheim tunneling.

$$J = \frac{q^3 V^2 m_0}{84\pi h \phi_\mathrm{B} m^*} \exp\left(-\frac{4(2m^*)^{0.5}\Phi_\mathrm{B}^{1.5}}{3hqV}\right) \tag{7.3}$$

where m^*: effective mass of carrier, k_B: Boltzmann constant, h: Planck constant, T: temperature, q: element charge, Φ_B: barrier height, V: applied voltage, m_0: mass of free electron. Though the injection mechanism has been discussed on the basis of the above mentioned theory through the temperature dependence of the current and so on [7], it would be necessary to consider some sort of inter gap states or another mechanism to lead high electric field locally at the interface in order to overcome the large barrier height estimated from the work function of the electrodes and the carrier transport level.

The reduction of the injection barrier is one of the key issues in OLEDs. In practical cases, the injection barrier is reduced through interfacial states produced artificially by inserting very thin buffer layer or doping electron acceptors/donors at the interface. The typical materials often used for buffer layer are shown in Fig. 7.6.

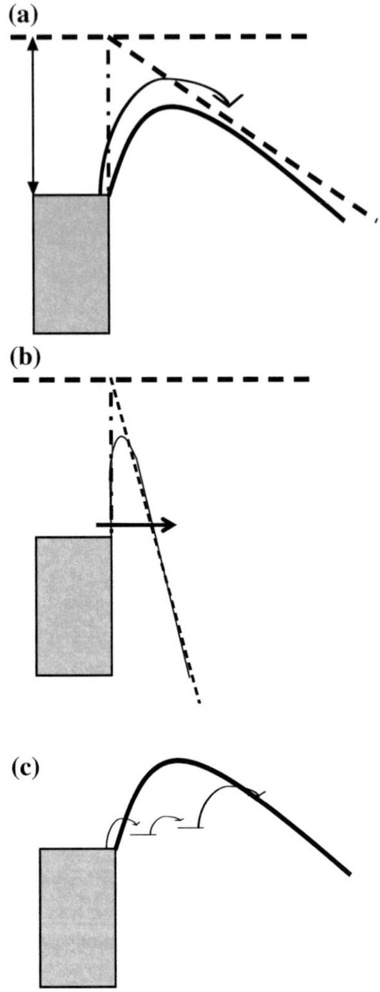

Fig. 7.5 Energy diagram at the interface. **a** Schottky thermal emission. **b** Flower-Norheim tunneling, **c** Injection through the interfacial state

7.3.2 Carrier Transportation

The injected charge carriers are transported in the amorphous transporting materials according to the hopping mechanism. The carrier mobility is typically below 10^{-1} cm^2/Vs even in the most advanced carrier transport materials [8] (Figs. 7.7, 7.8 and 7.9).

Even in the case that the injection barrier is negligible, the current is not proportional to the applied voltage at the device operation region because it is governed by the space charge formed in the organic film. It is called space charge limited current (SCLC) [9]. The current is proportional to the square of the voltage when density of the carrier trap is also negligible as expressed in Eq. 7.4.

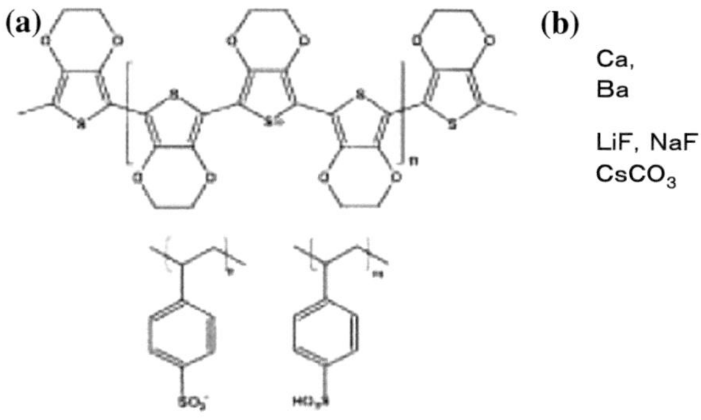

PEDOT:PSS

Fig. 7.6 Typical materials used for **a** anode buffer layer and **b** cathode buffer layer

$$J = \frac{9}{8}\epsilon_0\epsilon\mu_e\frac{V^2}{d^3} \tag{7.4}$$

where ϵ_0: vacuum permittivity, ϵ: dielectric constant, μ_e: carrier mobility, d: thickness of the film. In the low current region, the current is observed to be proportional to the applied voltage. In this ohmic region the intrinsic carriers caused by unintended impurities are responsible. At certain voltage, the amount of the injected carrier overtakes the intrinsic carrier. The current becomes proportional to V_2 after abrupt increase (Fig. 7.10).

When the trapped charge in the film forms space charge the current tends to be proportional to the higher order of magnitude of the voltage.

7.3.3 Carrier Recombination

The transported hole and electron are recombined on a molecule. The largest advantage of the stacked layer structure of both transporting materials should be to improve the carrier balance. The holes traveling through the HTM would be accumulated at the HTM/ETM interface due to the energy gap between HOMO levels of both transporting materials. The electron would also be accumulated in the same manner (Fig. 7.11). The probability of recombination can be maximized in this context.

Though the injection barrier height and carrier mobility are not equal for hole and electron, the charge balance factor can be improved when the each transporting material is stacked to build the effective block for both carriers at the interface.

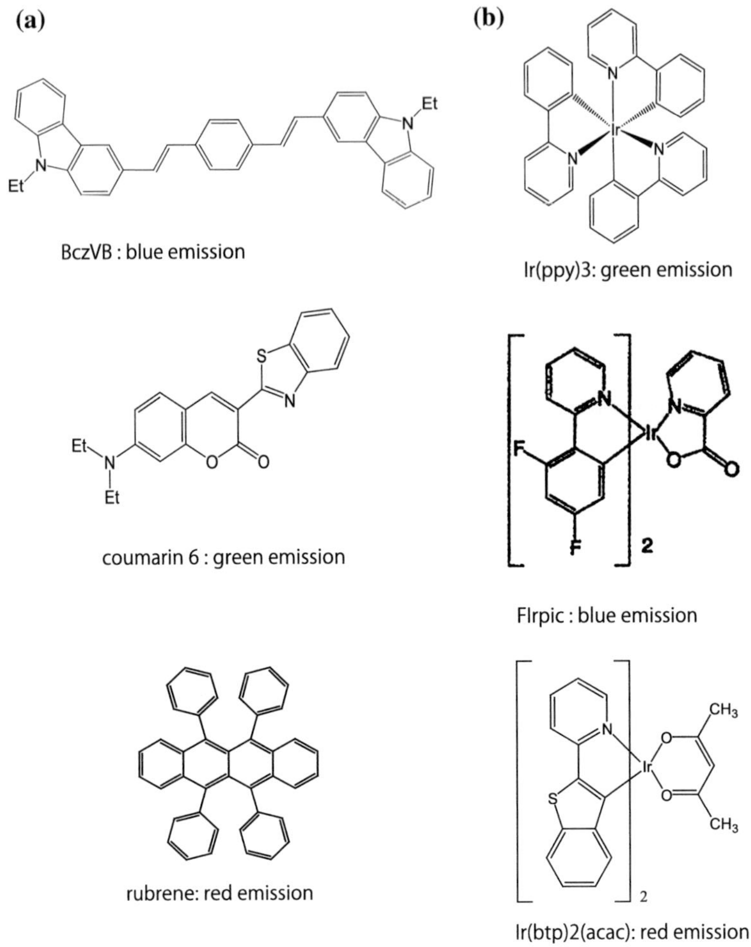

Fig. 7.7 Molecular structure of **a** fluorescence dyes and **b** phosphorescence dyes

7.4 Emissive Species

7.4.1 Energy Transfer

To improve the ϕ_f, it is quite effective to transfer the excitation energy from the recombination site to a luminance center having higher quantum yield of luminance. Small molecular dye is often doped in the emission layer of OLED. In this case, the luminance ascribed to the host material is quenched and that to the doped dye can be observed. Since the ϕ_f, of the dye is much higher than that of the host. The overall ϕ_f can be also getting higher when the energy transfer rate is effectively high. The excited energy transfer takes place through dipole-dipole interaction or electron exchange

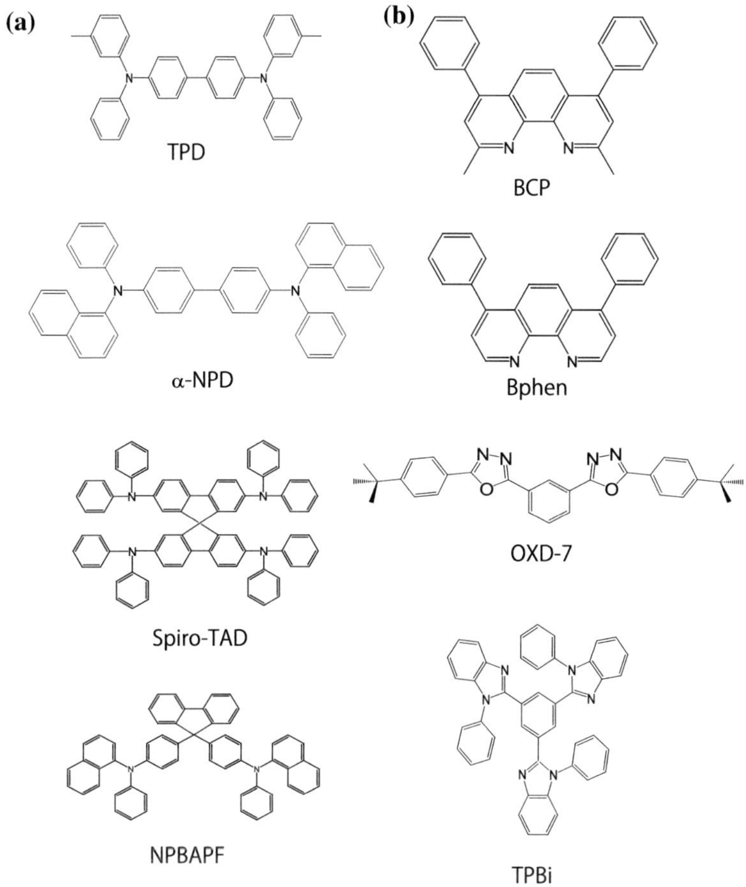

Fig. 7.8 Molecular structure of **a** hole transport materials and **b** electron transport materials

interaction between the host and the dopant. In the former, Förster mechanism [10], the transition dipole from the excited state to the ground state at the host molecule is resonant with the transition dipole from ground state to the excited state at the dopant, resulting in the energy transfer from the host to the dopant (Fig. 7.12a). The critical length of the energy transfer can be over a few nm, and, however, the both transitions between the excited states and the ground states have to be allowed in this mechanism. In the latter, Dexter mechanism [11], the excited energy is transferred by changing the electron at the excited state in the host with that at the ground state in the dopant. This type of transfer can take place in very short distance (Fig. 7.12b).

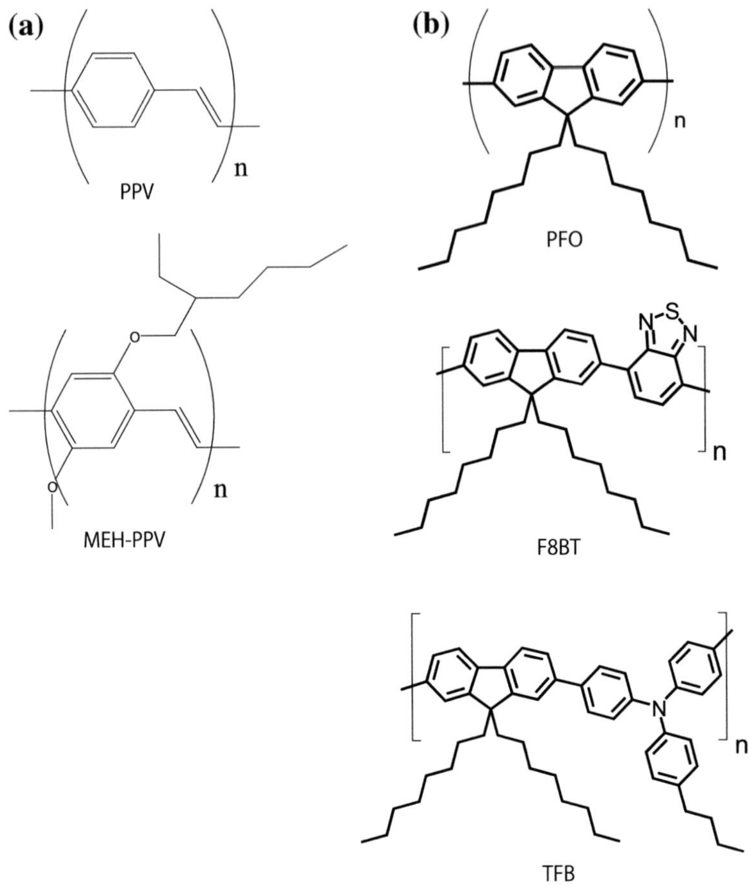

Fig. 7.9 Molecular structure of **a** poly(p-phenylenevinylene) derivatives and **b** polyfluorene derivatives

7.4.2 Fluorescence Dye

For the singlet exciton, the energy transfer can take place through Förster mechanism since the transition between the excited state and the ground state is allowed. The transfer rate constant is proportional to the overlap between the fluorescence spectra of the host molecule and the absorption spectra of the dopant. When the overlap is large enough and the quantum yield of the dopant is sufficiently high the critical length of the energy transfer can be as long as 10 nm. In such a case, the luminance of the host can be quenched completely in only 0.1 % of doping ratio and the ϕ_f of the doped film can be increased up to unity.

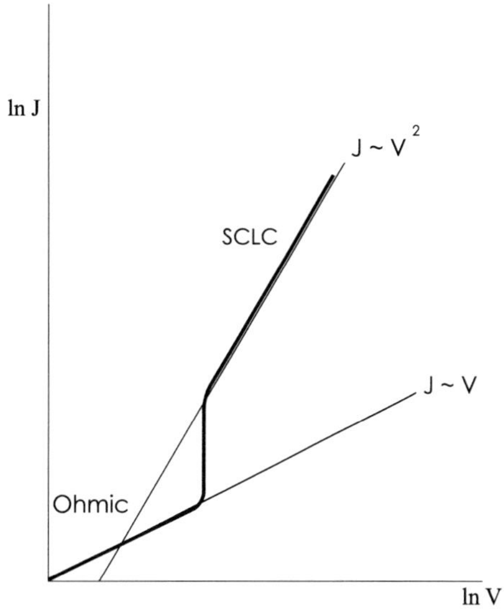

Fig. 7.10 Typical current density-voltage characteristics of organic carrier transport materials in SCLC

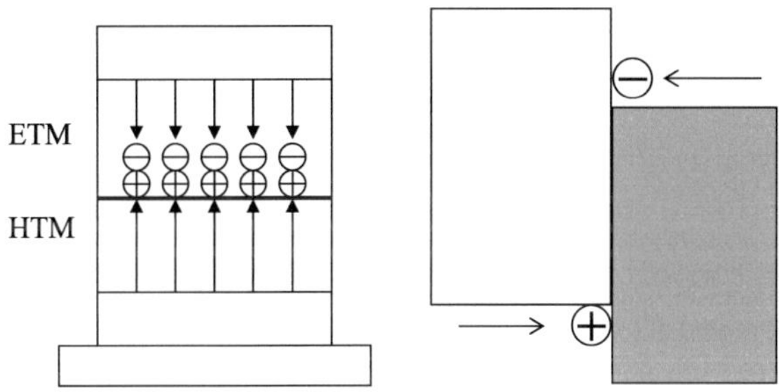

Fig. 7.11 Carrier confinement by the stacked layer structure

7.4.3 Phosphorescence Dye

As mentioned in Sect. 7.2, the decay process of the triplet exciton to the ground state is usually forbidden. Since 75 % of the exciton generated by the carrier recombination should be the triplet exciton, it is quite effective for the improvement in β that this transition is utilized for EL [12]. In ordinary organic materials the luminance based on the direct radiative decay of triplet excitation state to the ground state, phosphorescence, can be observed at quite low temperature to suppress thermal decay processes. However, some metal complexes having a heavy atom (Fig. 7.7b), such as Ir [13] and Pt [14] show the intense phosphorescence even at the room temperature. The strong internal heavy atom effect makes the direct radiative decay process and the exchange process between the singlet exciton and the triplet exciton allowed. The EQE of the OLEDs was drastically increased by employing Ir complex phosphorescence dye. The phosphorescence dyes are used as a dopant in a host

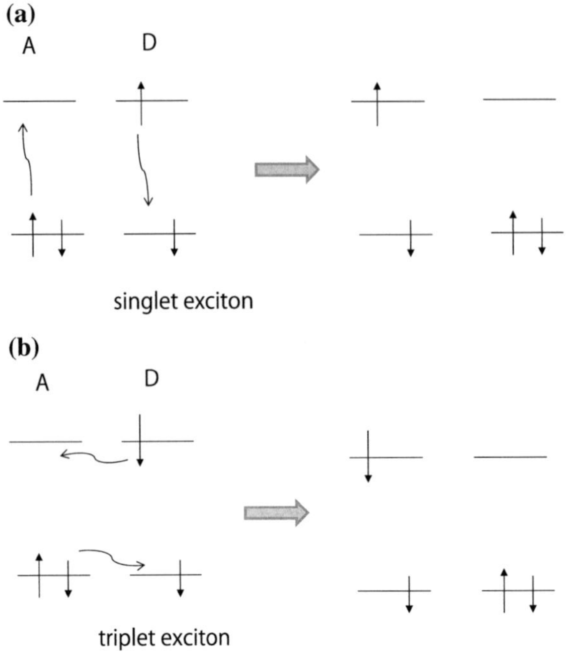

Fig. 7.12 Excitation energy transfer in **a** Förster mechanism and **b** Dexter mechanism

material with higher triplet level for the energy transfer from the host to the doped dye. The energy transfer takes place in Dexter mechanism with much shorter critical distance because the transition between the excited state and the ground state is forbidden in the host molecule though it is allowed in the phosphorescence dyes. For efficient energy transfer, therefore, the much higher doping ratio up to several percent is required than the fluorescence dye. Since the energy transfer between triplet excitons in the phosphorescence dyes can take place in Forster mechanism due to the allowed transition, the exciton can migrate among phosphorescence dyes at the high doping ratio.

7.5 Materials

7.5.1 Small Molecules

Huge number of hole transport materials have been developed so far. Most of them have an aromatic amine group. The aggregation of the molecules during the device operation is practically the most serious problem. When the temperature rises due to the Joule heat, a molecular becomes mobile above the glass transition temperature,

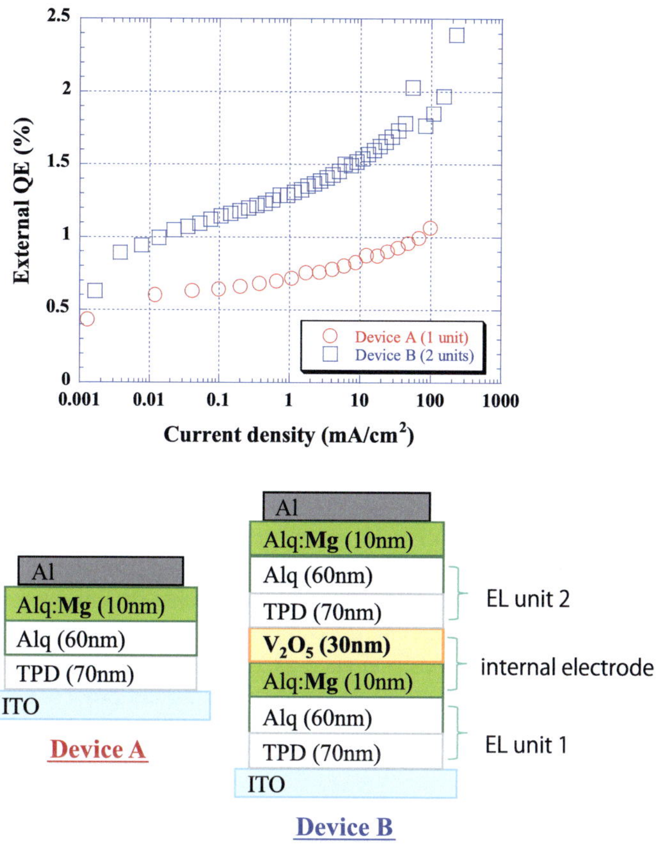

Fig. 7.13 Schematic illustration of single unit OLED (device A) and tandem OLED (device B), and the EQE characteristics of device A and B

Tg, and tends to aggregate. Therefore, the development policy of the hole transport materials have been to find molecules showing higher Tg and lower tendency for aggregation. It has been successful strategy to combine some triphenylamine groups into a molecule in this context. Many kinds of electron transport materials have also been developed. Metal quinoline complex, oxadiazole and phenanthroline derivatives are known for efficient backbones.

7.5.2 Polymers

Polymers can be made into thin film by simple solution process such as printing method and spin-coat technique which is potentially cost effective. On the other hand, it is hard to obtain extra pure materials because standard purification technique, vacuum sublimation, cannot be applied. Polymers used for OLEDs can be classified into two, pendant polymers and π-conjugated polymers. Pendant polymers have basically a non-conjugated backbone, mostly polyethylene backbone, and above-described carrier transport small molecules as side chains.

The π-conjugated polymers have a conjugated backbone. In the first report of polymer OLED poly(p-phenylenevinylene) was used [15]. Though many derivatives have been developed, the luminance color range which can be realized using this backbone is not sufficient enough. Nowadays most of π-conjugated polymers for OLEDs have a polyfluorene backbone. They are basically blue emitting polymer and their luminance color can be tuned in wide rage by another moieties copolymerized in the backbone.

7.6 Device Fabrication and Architecture

The fabrication method for OLEDs is classified into two, a dry process and a wet process. The former is based on vacuum evaporation and has advantages of minimized impurities using clean environment and of precise structure control such as multi-layered structure by sequential deposition and doping by coevaporation. The latter is based on solution deposition such as printing, spin-coating and spraying, and potentially low-cost and large scale production. The dry process is now widely employed for the commercial production of flat panel displays. However, the wet process is getting to attract much attention as a large area device like lightings is targeted on.

7.6.1 Tandem Structure

Tow or three EL units are piled and stacked with internal electrodes. The apparent luminance efficiency can be doubled or tripled because the same current density would be supplied to each EL unit though the doubled or tripled voltage is required [16]. The key point of this structure is the formation of the internal electrode. It can generate hole and electron and inject into the EL units on either side. It has a stacked layer structure of p-doped electron donor material/n-doped electron acceptor material (Fig. 7.13).

7.6.2 Improvement of the Coupling-Out

Finally coupling-out factor is discussed in this section. The generated photons in an organic molecule have to pass through the organic layer, the transparent electrode and the glass substrate into air. The refractive index of organic layer is relatively high,

1.5–1.7 at the visible light region. The coupling-out factor is estimated to be only 20 % by using simple Snell law and a slab model [17]. To increase this factor, the surface of the substrate was modified by attaching micro lenses, pasting microparticle scatterer or scratching a pattern.

7.7 Concluding Remarks

The EQE of OLEDs has been improved by blushing up each elementary process and increasing each factor. For the practical device application many other features than EQE has to be considered, such as color reproduction, device life time, production cost and so on. All features are now getting optimized by development of materials, device structure and fabrication process.

References

1. T. Tsutsui, MRS Bull. **22**, 39 (1997)
2. H. Helfrich, W.G. Schneider, Phys. Rev. Lett. **14**, 229 (1965)
3. P.S. Vincent, W.A. Barlow, R.A. Hann, G.G. Roberts, Thin Solid Films **94**, 171 (1982)
4. C.W. Tang, S.A. VanSlyke, Appl. Phys. Lett. **51**, 913 (1987)
5. H. Helfrich, W.G. Schneider, J. Chem. Phys. **44**, 2902 (1966)
6. G.G. Steetman, S. Banerjee, *Solid State Electronic Devices* (Prentice Hall, Upper Saddle River, 2000)
7. I.D. Parker, J. Appl. Phys. **75**, 1656 (1994)
8. M. Stolka, J.F. Yanus, D.M. Pai, J. Phys. Chem. **88**, 4707 (1984)
9. M.A. Lampert, Repot. Prog. Phys. **27**, 329 (1964)
10. T. Förster, Ann. Phys. **2**, 55 (1948)
11. D.L. Dexter, J. Chem. Phys. **21**, 836 (1953)
12. T. Tsutsui, C. Adachi, S. Saito, *Electroluminescence in Organic Thin Films, Photochemical Processes in Organized Molecular Systems* (Elsevier Sci. Pub., Amsterdam, 1991)
13. M.A. Bald, S. Lamansky, P.E. Burrows, M.E. Thompson, S.R. Forrest, Appl. Phys. Lett. **75**, 4 (1999)
14. D.F. O' Brien, M.A. Bald, M.E. Thompson, S.R. Forrest, Appl. Phys. Lett. **74**, 442 (1999)
15. J.H. Burroughes, D.D. Bradley, A.R. Brown, R.N. Marks, K. Mackay, R.H. Friend, P.L. Burns, A.B. Holmes, Nature **347**, 539 (1990)
16. L.S. Liao, K.P. Klubek, C.W. Tang, Appl. Phys. Lett. **84**, 2 (2004)
17. J.-S. Kim, P.K.H. Ho, N.C. Greenham, R.H. Friend, J. Appl. Phys. **88**, 1073 (2000)

Chapter 8
Magneto-Optical (MO) Characterization

Shinya Koshihara

Abstract It is well known that the anisotropy in condensed matters becomes the origin of linear and circular birefringences. In this section, the effect of magnetic field on optical properties, i.e. magneto-optical (MO) effects are introduced. Lately, the dynamical properties of MO property is becoming important, because change in MO property is deeply relating with the dynamical behavior of spin in advanced materials and it plays the key role in the growing field of 'Spintronics'. Here, we also introduce the modern techniques for probing the ultrafast change in MO property.

8.1 What is the Magneto-Optical (MO) Effect?

It is well known that the anisotropy in condensed matters becomes the origin of linear and circular birefringences. For example, the external electric field induces the anisotropy in various crystalline medias and result in birefringence even though they are originally isotropic. Similar effect can be expected for the magnetic field $\boldsymbol{B}$. The effect of a static field $\boldsymbol{B}$ whose frequency is much smaller than the optical frequency of the light wave can be given by a polarization term as follows:

$$P_i(\omega) = \epsilon_0 \sum_{jk} \gamma_{ijk} R_j(\omega) B_k + \text{higher order terms of } \boldsymbol{B}_k \tag{8.1}$$

The first term in this relation results into the Faraday effect and following ones into the magnetic Kerr and Cotton-Mouton effects (for detailed explanation and discussion, please refer to Refs. [1–3].

S. Koshihara (✉)
Department of Materials Science, Graduate School of Science and Engineering, Tokyo Institute of Technology, 2-12-1-H61 Oh-okayama, Meguro-ku, Tokyo, 152-8551 , Japan
e-mail: skoshi@cms.titech.ac.jp

Y. Aoyagi and K. Kajikawa (eds.), *Optical Properties of Advanced Materials*, Springer Series in Materials Science 168, DOI: 10.1007/978-3-642-33527-3_8,

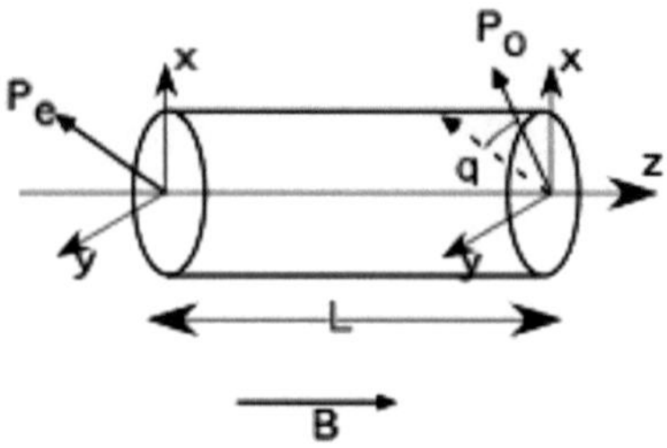

Fig. 8.1 Polarization rotation by Faraday effect

(a) Faraday effect
In case that an optically isotropic materials is put into a magnetic field $\boldsymbol{B}$, a light wave with linear polarization propagating in the direction of $\boldsymbol{B}$ has its polarization plane turn of an angle θ (see Fig. 8.1) given by the following relation:

$$\theta = VBL \tag{8.2}$$

In this expression, V represents a constant called Verdet's constant and L is the optical path length. This magnetic effect can be explained by the expression of the polarization induced by the magnetic field. Let's consider the case that the magnetic field is applied parallel to the z axis. The polarization induced along the i axis is given by

$$P_i(\omega) = \epsilon_o \gamma_{ijk} E_j(\omega) B_z \quad (k \neq z) \tag{8.3}$$

Tensor $[\gamma]$ is imaginary to fulfill the energy conservation law in non-absorbent medium. As a result, the effective relative permittivity tensor is given by:

$$[\bar{\epsilon}] = [\epsilon] + \begin{bmatrix} 0 & -i\gamma B & 0 \\ i\gamma B & 0 & 0 \\ 0 & 0 & 0 \end{bmatrix} \tag{8.4}$$

In case of an optically isotropic materials without B, the effective permittivity tensor can be simplified as follows [1]:

$$[\bar{\epsilon}] = \begin{bmatrix} \epsilon & -i\gamma B & 0 \\ i\gamma B & \epsilon & 0 \\ 0 & 0 & \epsilon \end{bmatrix} \tag{8.5}$$

The relation between **D** and **E** gives the following relation,

$$D = \epsilon_0[\bar{\epsilon}]E = \epsilon_0([\epsilon] + i\gamma B \times E) \tag{8.6}$$

In this case, the eigenstates of a plane wave propagating in the direction of the magnetic field are the right- and left-hand circular polarized lights (denoted as σ^+ and σ^-) (see Fig. 8.2). The refractive indices for each eigenstates are denoted as n_r and n_l and they are given by following equations:

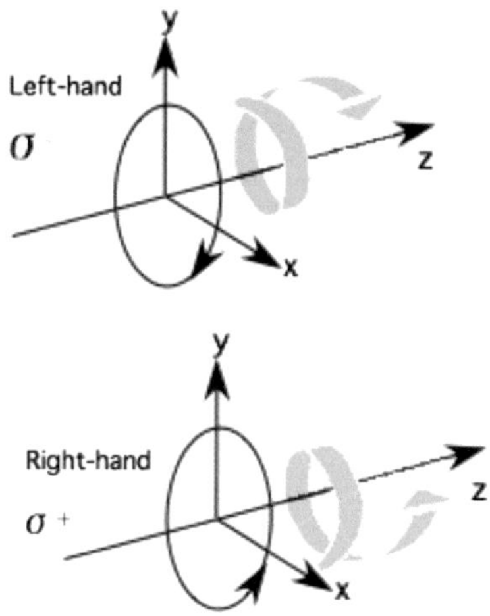

Fig. 8.2 Rotation of polarization direction in *right*- and *left-hand* circular polarized lights (denoted as σ^+ and σ^-)

$$n_l = n_0 + \gamma B/2n_0 \qquad n_r = n_0 - \gamma B/2n_0 \tag{8.7}$$

Here, n_0 is the index of the isotropic medium. As a result, the magnetic field induced circular birefringence is given by

$$\Delta n_c = \gamma B/n_0 \tag{8.8}$$

Utilizing this relation, the rotation angle θ and Verdet's constant V are obtained as follows:

$$\theta = \Delta n_c k_0 L/2 = \pi \Delta n_c L/\lambda = (\pi\gamma/\lambda n_0)BL \tag{8.9}$$

thus,

$$V = \pi\gamma/\lambda n_0 \tag{8.10}$$

As mentioned before, L denotes the length of transversed materials. Verdet's constant for usual materials are relatively small, and thus intense magnetic fields is necessary to in order to obtain a significant rotation of the polarization. However, in some advanced materials, large V value can be achieve by virtue of magnetic interaction and rather strong spin-orbit coupling as discussed in the following section.

In addition, Faraday effect is utilized for various optical devices. One typical example is an intensity modulator. Figure 8.3 shows the schematic viewgraph of the intensity modulator [1, 2]. When the rotation angle of the polarization direction supplied by Faraday effect is equal to θ and the angle between the polarizer and analyzer is equal to α, the transmitted intensity is proportional to $\cos^2(\theta - \alpha)$. If the $\alpha = \pi/4$, the total intensity modulation can be achieved by the variation of B between $-B$ and $+B$. Another important example is an optical isolator. The Faraday rotation is a non-reciprocal effect, thus the reverse trip through the medium provokes a rotation of the same magnitude but of opposite sense regarding to the direction of

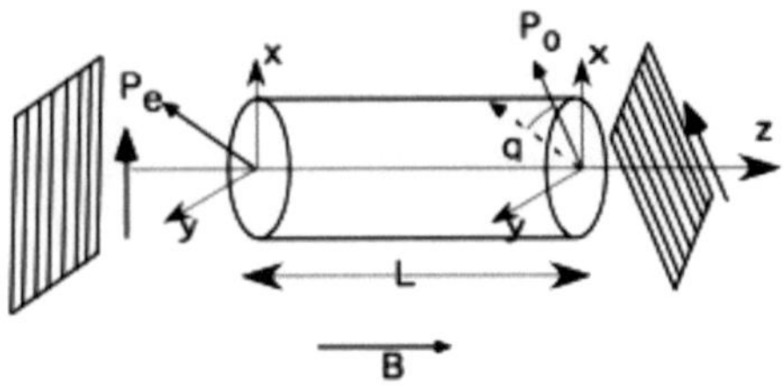

Fig. 8.3 Schematics of intensity modulator utilizing Faraday effect

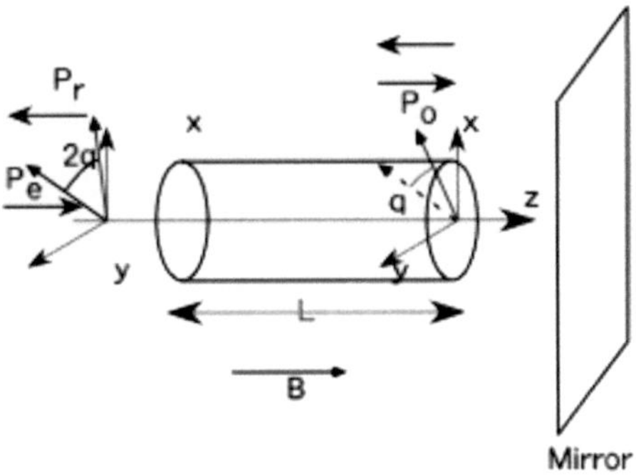

Fig. 8.4 Principle of a Faraday rotator

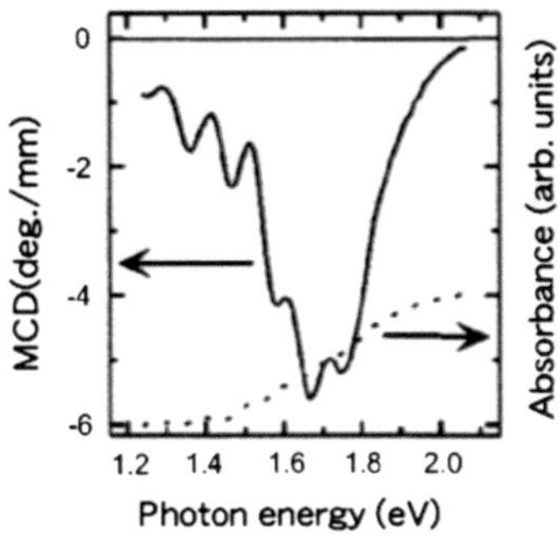

Fig. 8.5 MCD (*solid lines*) and absorption (*dotted lines*) spectra observed for ferromagnetic (Ga,Mn)As thin film at 5.0 K and magnetic field at 1.0 T [4]

propagation (see Fig. 8.4). Let's consider the reflected light by mirror shown in this figure. The total rotation angle (θ_t) after a double trip through the Faraday medium becomes the twice of the one-way trip:

$$\theta' = 2\theta = 2VBL \tag{8.11}$$

If we set $VBL = \pi/4$, the reflected light polarization is orthogonal to the irradiated one. The association of polarizer and this device construct an optical isolator which is essential device for optical high speed communication [2]. In this section, we have discussed the perfect transparent materials for convenience of discussion. However, in absorptive media, the magnetic circular dichroism (MCD), i.e. the difference in the absorption coefficient for left (σ^+) and right (σ^-) hand circular polarized light

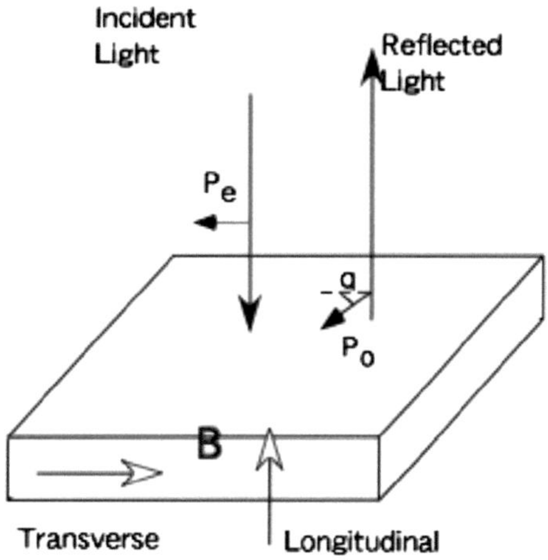

Fig. 8.6 Schematic of the magnetic Kerr effect in a material subject to a field *B* [1]

($\Delta\alpha_c = \Delta\alpha_+ - \Delta\alpha_-$), will be more dominant (see Fig. 8.5) [4, 5]. Anyway, origin of the Faraday rotation and MCD effects are difference in the electronic response of the materials to two kinds (left and right hand) of circular polarized light waves. The relation between MCD and electronic structure of advanced materials will be discussed in detail in the following section.

(b) Magneto-Kerr effect

When the linear polarized light is reflected from the surface of the materials under magnetic field, the polarization direction of the reflected light is rotated with angle θ relatively to the irradiated one (see Fig. 8.6). This effect is named as the magneto-Kerr effect. In case that the magnetic field $\boldsymbol{B}$ is applied to the materials in a plane parallel to the surface, it is called as transverse effect. When it is perpendicular to the surface, it is called as longitudinal effect. This rotation of the polarization direction in reflected light is due to the difference in the reflection coefficients of the left- and right-hand circular polarized light, since the index of the refracting surface depends on the incident eigenstate. Kerr effect is very small but it is the basis upon which magneto-optical (MO) disks are working as shown in Fig. 8.7 [1, 2]. In case of record, MO disk is located under the magnetic field B whose amplitude is smaller than the coercive field at room temperature. When zone A is heated beyond Curie temperature by laser irradiation, the coercive field for this region becomes very small (B_c) and the magnetization of this region is aligned parallel to the direction of $\boldsymbol{B}$ (see Fig. 8.7a). When we read the recorded data, the linearly polarized probe light is irradiated on the surface of the MO disk and the direction of magnetization in every recording zone (see Fig. 8.7b) is detected by Kerr rotation effect of the reflected light. To erase the recorded data, the disk is heated up by laser irradiation under the magnetic field parallel to the initial direction of the magnetization as shown in Fig. 8.7c.

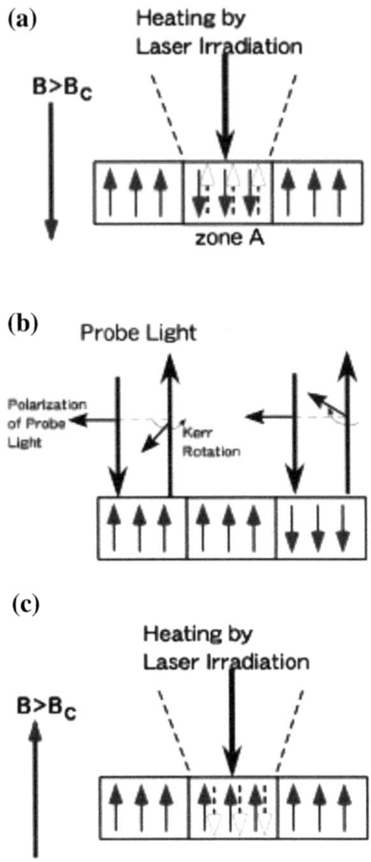

Fig. 8.7 Principle of the process for data storage in the magneto-optical disks. B_c denote the coercive magnetic field at the ambient pressure under the laser heating condition [1]. **a** Recording process. **b** Reading process. **c** Erasing process

8.1.1 What is the Origin of MO Effect in Advanced Materials?

As discussed in the previous section, in ordinal case, MO effect is rather small. Therefore it is very difficult work to achieve the integration of MO device (Faraday isolator etc.) into the semiconductor and opto-electronic (OE) devices. However, recently, new class of materials which are named as II-VI and/or III-V based diluted magnetic semiconductors (DMS) have been developed to solve this problem. In this class of materials, by virtue of magnetic interaction mediated by carriers, rather strong Spin-Orbital (LS) coupling interaction and large optical gap energy in based semiconductors, very large circular birefringence (Δn_c) responsible to external weak magnetic field has been achieved in near Infrared and Visible regions [5–7]. Therefore, the application of this class of materials to the compact OE devices is becoming a realistic target. In the following part of this section, the discussion will be focused on III-V based DMS materials ((Ga,Mn)As) with zinc blende (ZB) type structure as typical advanced materials for the application to MO devices. In (Ga,Mn)As, part of Ga (about 0.1–5 %) atoms are substituted by Mn^{2+} ions which work as magnetic centers based on d5 electrons with high-spin configuration.

As is well known, the magneto-optical properties in ZB type II-VI-based DMS such as (Cd,Mn)Te can be sufficiently explained by the Zeeman splitting effect of the band-edge state based on the exchange interaction between localized d-electrons of Mn^{2+} ions and itinerant band electrons and/or holes (s,p-d exchange interaction) [5, 6]. Here, the MO properties of (Ga,Mn)As will be explained based on the similar model proposed by Szczytko et al. [8] including the effect of disorder and change in the Fermi energy level. Following this method, the absorption coefficient which includes the effect of both the disorder effect and the changing Fermi energy is defined as follows:

$$\alpha(\hbar\omega) = A'\frac{m_{\mathrm{v}}a^2}{\hbar\omega}\int_0^{k_{\mathrm{c}}^{\max}} dk_{\mathrm{c}}k_{\mathrm{c}}\left[\exp\left(-\frac{a^2}{2}(k_{\mathrm{c}}-k_{\mathrm{v}})^2\right) - \exp\left(-\frac{a^2}{2}(k_{\mathrm{c}}+k_{\mathrm{v}})^2\right)\right]; \tag{8.12}$$

here,

$$k_{\mathrm{v}} = \sqrt{\frac{2m_{\mathrm{v}}}{\hbar^2}\left(\hbar\omega - E_{\mathrm{g}} - \frac{\hbar^2 k_{\mathrm{c}}^2}{2m_{\mathrm{c}}}\right)} \tag{8.13}$$

$$k_{\mathrm{c}}^{\max} = \sqrt{\frac{2m_{\mathrm{c}}}{\hbar^2}(\hbar\omega - E_{\mathrm{g}} - F(n_{\mathrm{p}})} \tag{8.14}$$

and

$$A' = \frac{2A^2(\hbar\omega)}{(2\pi)^{5/2}\hbar^2} \tag{8.15}$$

$m_{\mathrm{c(v)}}$ and $k_{\mathrm{c(v)}}$ are the effective mass and the wave vector of the quasimomentum for the conduction (valence) band, and E_{g} and F are the band gap energy and the Fermi energy, respectively. n_{p} is the hole-carrier density. The value 'a' stands for the degree of the expanse of the electric wave function. The Fermi energy can be estimated from n_{p}.

To calculate the MCD ($\Delta\alpha_{\mathrm{c}}$) spectrum which is defined as a difference between the absorption coefficients for circularly polarized (σ^+ and σ^-) light (see Fig. 8.8), the theoretical estimation of the absorption spectra for $\sigma^{\pm}$ light (denoted as $\alpha_{\pm}(\hbar\omega)$) under a magnetic field is necessary. In the case of III-V semiconductor crystals with ZB type structure, $\alpha_{\pm}(\hbar\omega)$ is given by the sum of the absorption coefficients originated from the electronic transitions from three valence bands, i.e., heavy hole, light hole, and spin-orbit split bands, to the conduction band, as shown in Fig. 8.9. For the estimation of these components in $\alpha_{\pm}(\hbar\omega)$ based on Eq. 8.12, the E_{g} and $F(n_{\mathrm{p}})$ values in Eqs. 8.13 and 8.14 have to be changed into effective values due to spin splitting under the magnetic field in the electronic band structure [8]. The effective E_{g} values for three components, i.e., the transitions to the conduction band from the

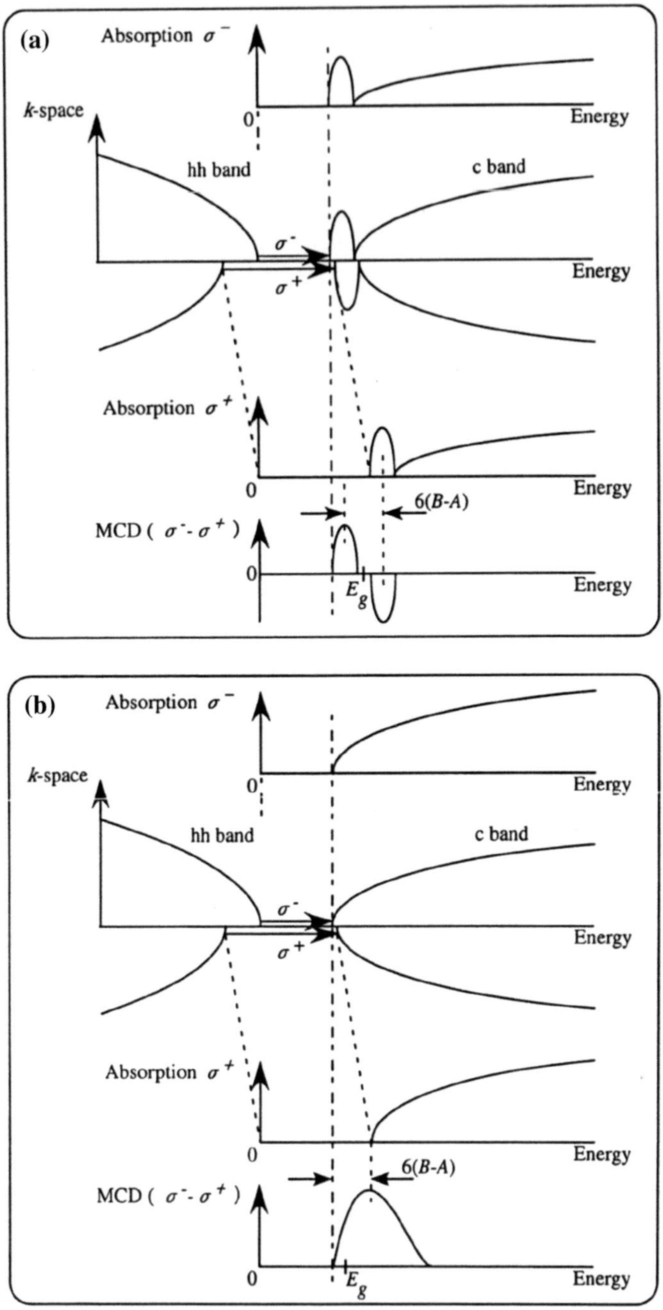

Fig. 8.8 The process for the calculation of MCD ($\Delta\alpha c$) spectrum which is defined as a difference between the absorption coefficients for circularly polarized (σ^+ and σ^-) light. **a** and **b** denote the case for excitonic absorption and band edge absorption, respectively. MCD can be obtained based on the theoretical estimation of the absorption spectra for $\sigma^{\pm}$ light under a magnetic field

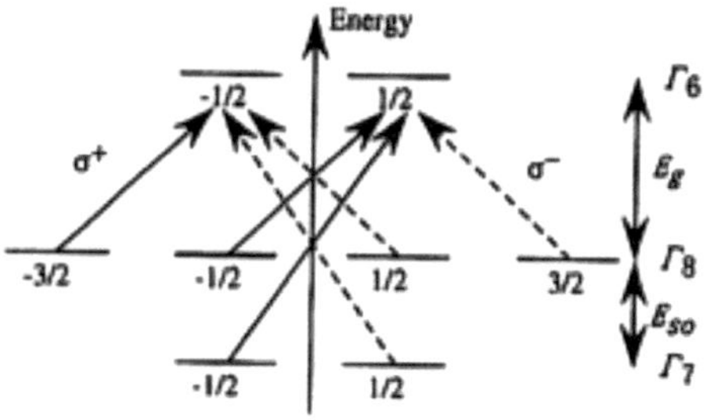

Fig. 8.9 A Schematic representation of band splittings in a Zinc-blende type semiconductors. Optical transitions in allowed Faraday configuration for σ^+ and σ^- light are denoted by *solid* and *dashed arrows*, respectively

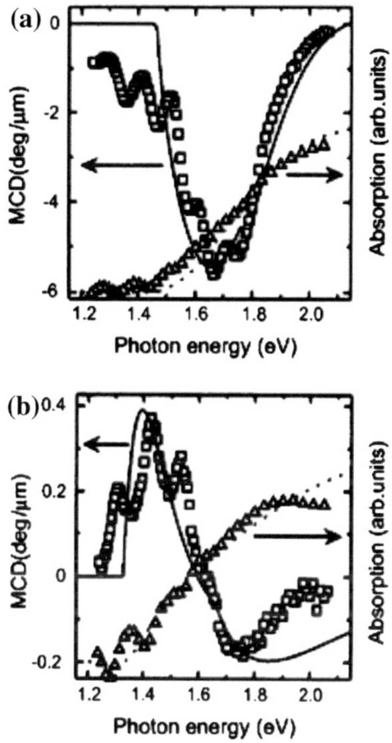

Fig. 8.10 Observed (*square* and *triangular plots*) and calculated (*solid* and *dotted lines*) MCD and absorption spectra for (Ga,Mn)As thin films with hole carrier densities of $n_p = 1 \times 10^{19}$ (**a**) and $1 \times 10^{18}\,\mathrm{cm}^{-3}$ (**b**). As for absorption spectra, the results of theoretical calculation for zero magnetic field are indicated [4]

heavy hole ($E_{\mathrm{hh-c}}$), light hole ($E_{\mathrm{lh-c}}$), and spin-orbit split ($E_{\mathrm{so-c}}$) bands in σ^+ and σ^- circularly polarized light are given as follows: For σ^+,

$$E_{\mathrm{hh-c},\sigma^+} = E_{\mathrm{g}} + 3b - 3a, \tag{8.16}$$

$$E_{\mathrm{lh-c},\sigma^+} = E_{\mathrm{g}} + b + 3a, \tag{8.17}$$

$$E_{\mathrm{so-c},\sigma^+} = E_{\mathrm{g}} - b + 3a, \tag{8.18}$$

and for σ^-,

$$E_{\mathrm{hh-c},\sigma^-} = E_{\mathrm{g}} - 3b + 3a, \tag{8.19}$$

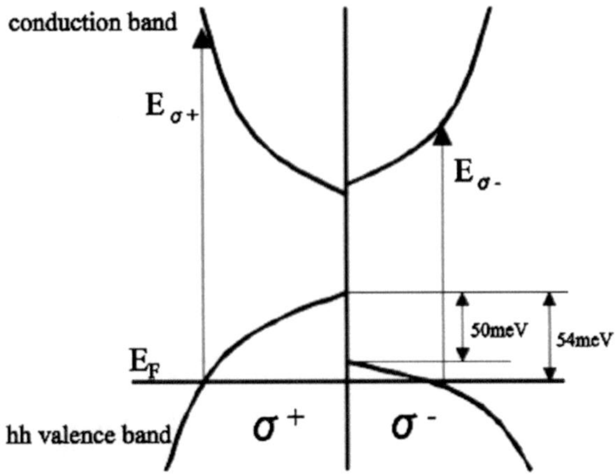

Fig. 8.11 Schematic band structure corresponding to ferromagnetic (Ga,Mn)As at a magnetic field of 1 T (only the conduction and heavy hole bands are indicated) [4]

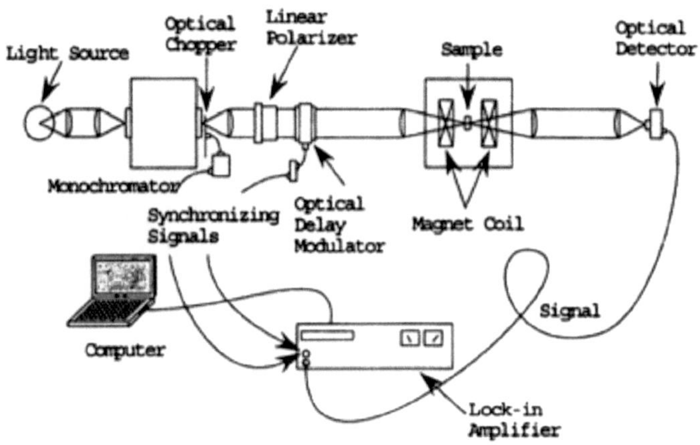

Fig. 8.12 Experimental setup for the measurement of MCD spectrum under various magnetic field

$$E_{\mathrm{lh-c},\sigma^-} = E_{\mathrm{g}} - b - 3a, \tag{8.20}$$

$$E_{\mathrm{so-c},\sigma^-} = E_{\mathrm{g}} + b - 3a, \tag{8.21}$$

where

$$a = \frac{1}{6} N_0 \alpha x \langle -S \rangle \tag{8.22}$$

$$b = \frac{1}{6} N_0 \beta x \langle -S \rangle \tag{8.23}$$

$N_o\alpha$ and $N_o\beta$ are the exchange integrals for the conduction (s-d) and valence (p-d) bands, and $\langle S \rangle$ and x are the thermodynamic average of Mn spin and the concentration of Mn, respectively. The solid and dashed lines in Fig. 8.10 shows the hole-density (n_p) dependence of the calculated MCD and the normal absorption spectra based on this method and properly estimated physical parameters [4]. The spectra for the hole densities of $n_\mathrm{p} = 1 \times 10^{19}$ and $1 \times 10^{18}\,\mathrm{cm}^{-3}$ are plotted in Fig. 8.10a, b, respectively, as typical examples, and they reproduce well the experimentally observed MCD (open squares) and absorption (open triangles) spectra of (Ga,Mn)As samples with corresponding carrier densities. The schematic band structure under a magnetic field of 1 T with the estimated spin splitting value and Fermi energy level for sample with $n_\mathrm{p} = 1 \times 10^{19}\,\mathrm{cm}^{-3}$ is summarized in Fig. 8.11 [4]. Based on these considerations, we conclude that the origin of the observed large MCD in DMS can be attributed to the ferromagnetic coupling among Mn ions and rather strong LS coupling. Because these two effects result in large $\langle S \rangle$ value and CD spectra strongly coupled with the direction of magnetic moment.

8.1.2 How to Measure the Magneto-Optical (MO) Characterization

The measurement of MO effect was difficult work, because MO effect is very small and the various optical components to control the polarization of light are necessary for experimental estimation. However, recently, highly sensitive detection of MO effect based on the polarization modulation technique can be utilized even in ordinary laboratory combining strong (<10T) superconducting magnet. For example, in Fig. 8.12, the experimental set up for measurement of MCD spectra are indicated. The MCD spectrum which is given as the difference between absorption coefficients for σ^+ and σ^- light can be estimated from the difference in transmitted light intensities for both polarized light. The discrepancy in the intensity of transmitted light depending on the light polarization can be highly sensitively detected based on the lock-in detection technique synchronized to the polarization modulation frequency. The results plotted in Fig. 8.10 are typical examples.

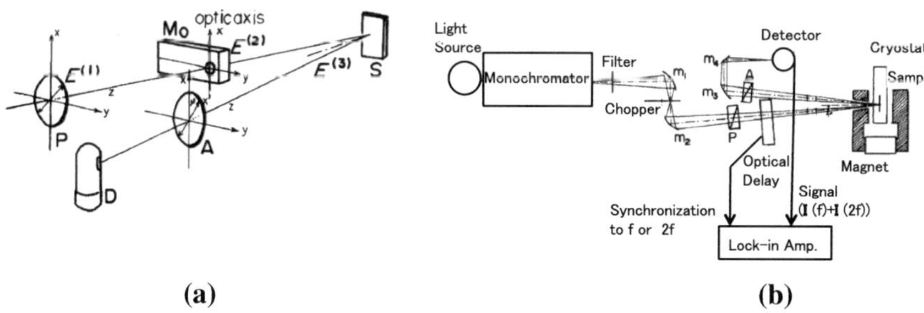

Fig. 8.13 **a** Schematic diagram showing the essentials of the measurement for magneto-optical Kerr effect [9]. **b** Detailed experimental set up [9]

In addition, utilizing the expansion of this technique, various MO properties can be observed as discussed by Sato [9]. Figure 8.13a, b shows the experimental set up for various MO properties in Kerr configuration. The detected signals synchronized to the modulation frequency f ($\mathbf{I}(f)$) and twice of it $2f$ ($\mathbf{I}(2f)$) are directly relating to Kerr rotation angle ϕ and Kerr ellipticity η as follows:

$$I(f) \propto \Delta R J_1(\delta_0) \tag{8.24}$$

$$I(2f) \propto \Delta R J_2(\delta_0) \sin(\Delta\theta + 2\phi) \tag{8.25}$$

$$\phi = -\frac{1}{2}\Delta\theta, \eta = \frac{1}{4}\left(\frac{\delta R}{R}\right) \tag{8.26}$$

Here, the optical retardation is modulated with frequency f:

$$\delta = \delta_0 \sin(2\pi f t) \tag{8.27}$$

R is the reflectivity of the material surface. In addition, ΔR and $\Delta\theta$ are differential reflectivity and differential rotation angle between σ^+ and σ^- light. Figure 8.14 shows obtained MO properties of Gd-Co film based on this method[9]. As for experimental technique for dynamical change in MO properties will be discussed in the following section.

8.2 Photoluminescence

Emission of radiation is the inverse of the absorption process. Most important condition for emission is that the system is not at equilibrium, i.e. system is excited by some external stimulations. The light emission process is generally named as luminescence. Excitation by an electric current results in electro-luminescence (EL) which is important phenomena for display devices. Emission induced by optical excitation via absorption and by electron beam are named as photo-luminescence and cathodo-luminescence, respectively. In this section, we focus on the photo-luminescence phenomena.

Figure 8.15 shows ordinary measurement system for photoluminescence and its excitation spectra. In this figure, for the measurement of excitation spectra with high resolution, the monochromator combined with photomultiplier is utilized for luminescence detection. Quite recently, as a result of the development of highly sensitive CCD detector, it is becoming possible to observe the multi-channel detection of luminescence spectrum utilizing polychromator coupled with CCD camera.

The photoluminescence process includes the various relaxation processes from photo-excited species. Therefore, the spectral shape, time dependence of it and time-profile of luminescence intensity after pulsed photo-excitation are essential information for the study of excited state dynamics. Figure 8.16 shows the time-resolved

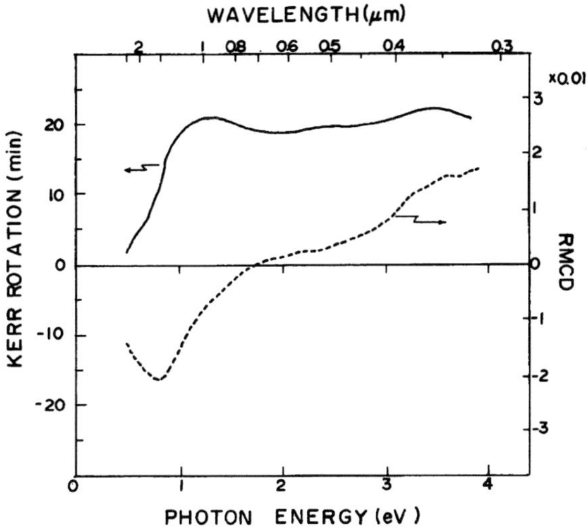

Fig. 8.14 Magneto-optical spectra of RF-sputtered amorphous Gd-Co film. *Solid line* shows Kerr rotation and *dotted* one denotes reflectance magneto-circular dichroism (RMCD) [9]

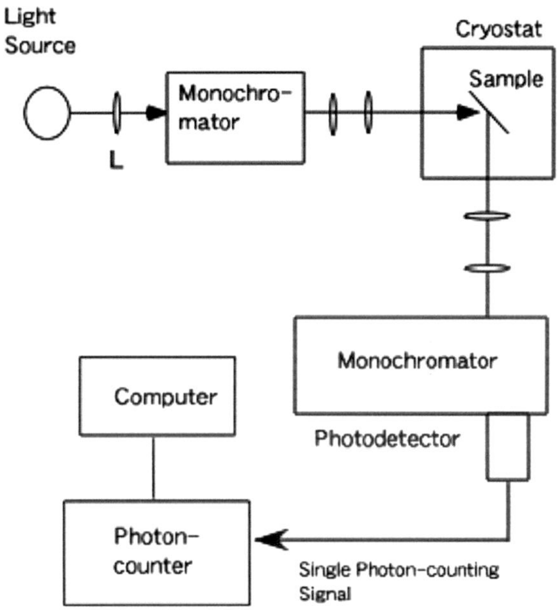

Fig. 8.15 Experimental setup for the measurement of photoluminescence spectrum and excitation spectrum of it

luminescence measurement system combined with time-resolved photon counting system (with the resolution about 50 ps). In addition, ultra-fast (less than 1 ps) time-resolution for photoluminescence can be achieved by utilizing streak camera system (Fig. 8.17) [10] and up-conversion method (Fig. 8.18) [11]. In Fig. 8.19, the temperature dependence of the time-resolved photoluminescence spectra for (Cd,Mn)Te are plotted as typical example [12]. Photoinjected species in (Cd,Mn)Te, i.e. exciton, is energetically relaxed as a result of the magnetic polarization. In other words, in this

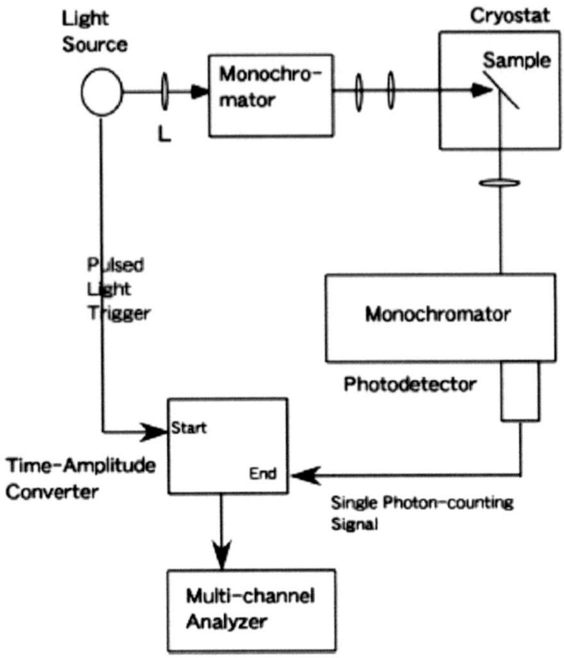

Fig. 8.16 Experimental setup for the measurement of time-resolved luminescence spectrum

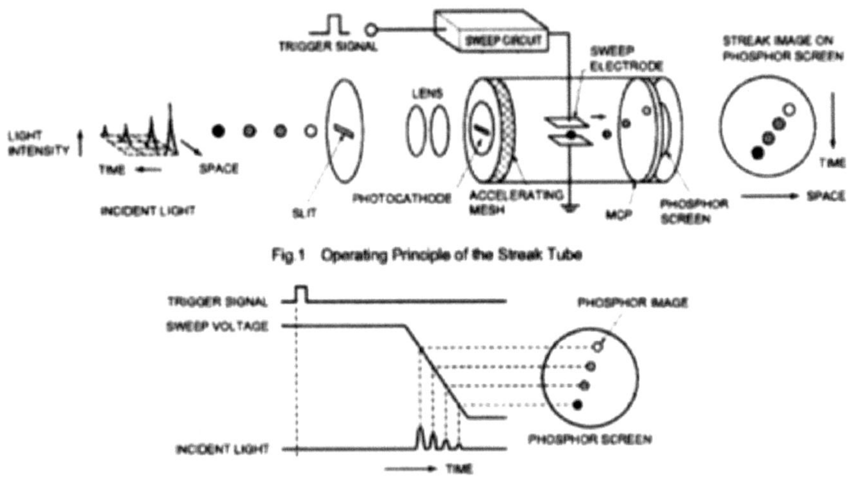

Fig. 8.17 Principle for the time-resolved luminescence measurement by streak camera [10]

class of advanced materials, magnetic polaron is formed in the relaxation process of exciton and it is directly reflected in the shift of the maximum energy for luminescence [5, 7]. In addition, external magnetic field is very effective for the shift in luminescent energy as shown in Fig. 8.20 [13].

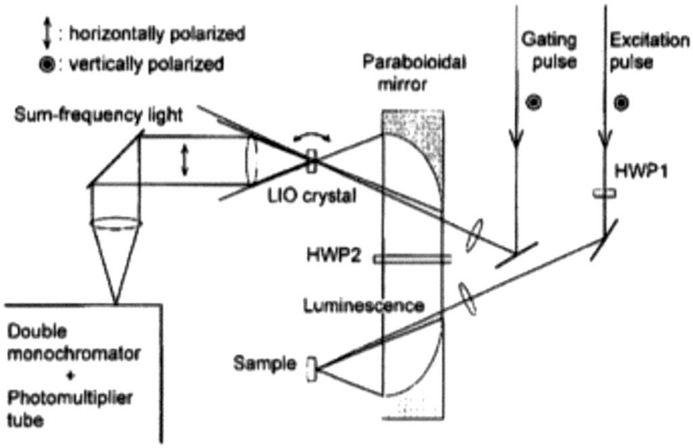

Fig. 8.18 Experimental setup for the femto-second time-resolved luminescence measurement. The luminescence of the sample is collected and focused onto a nonlinear optical crystal ($LiIO_3$ (LIO), thickness 0.5 or 1 mm) by off-axis paraboloidal mirrors. Another beam of the Ti:sapphire laser (gating pulse in the figure) is suitably delayed and focused onto the same nonlinear crystal. The time-resolved luminescence is detected as sum frequency photons generated from the luminescence and the gating laser pulse. The sum frequency light is collected and focused into a double-grating monochromator, and is detected by a photomultiplier tube with a photon counter. The time evolution is measured by varying the time delay of the gating pulse with respect to the sample excitation pulse [11]

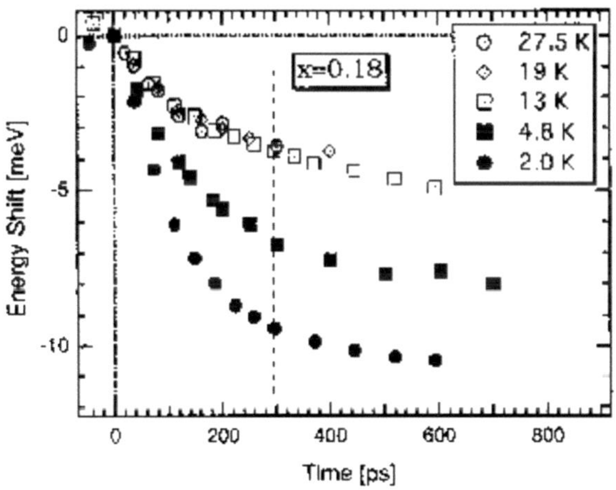

Fig. 8.19 Time evolution of the photo-luminescence energy shift for various temperatures observed for $Cd_{0.82}Mn_{0.18}Te$. The *vertical dotted line* indicates the time at which localization energies were determined [12]

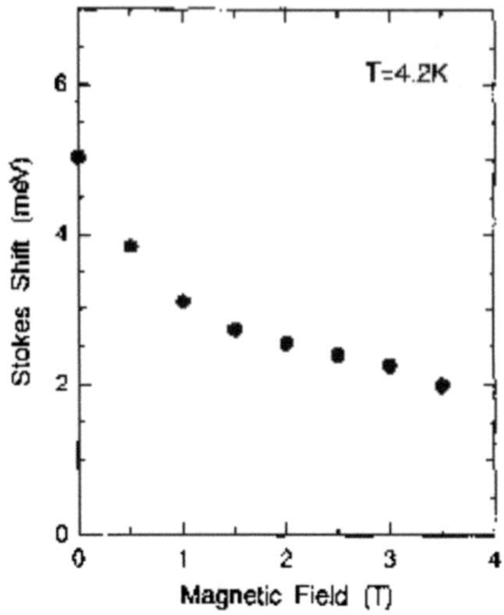

Fig. 8.20 Magnetic field dependence of Stokes shifts energy observed in photo-luminescence of the $Cd_{0.95}Mn_{0.05}$Te-ZnTe QW [13]

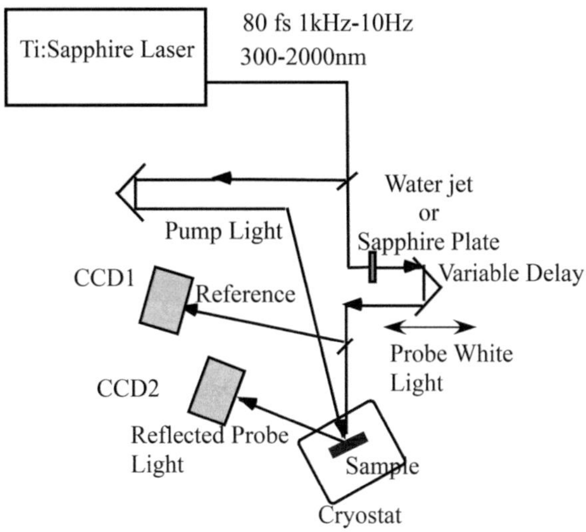

Fig. 8.21 Experimental setup for the measurement of time-resolved reflectance measurement with 150 femto-second resolution

8.2.1 Time-Dependent Method

As for time-resolved measurement method for photoluminescence has been treated in the previous section. Therefore, in this section, we will focus on time-resolved method for probing dynamical changes in optical and magneto-optical (MO) properties.

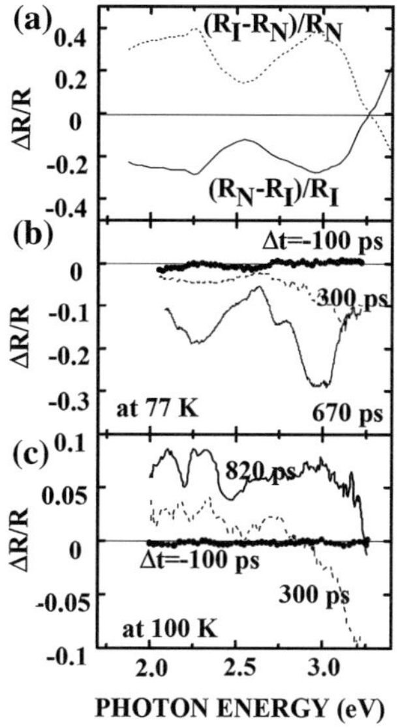

Fig. 8.22 **a** Calculated differential spectra for ionic to neutral phase transition $((R(N) - R(I))/R(I)$: *solid line*) and neutral to ionic transition $((R(I) - R(N))/R(N)$: *dashed line*). Here, $R(N)$ and $R(I)$ are the typical reflectance spectra in the N- and I-phases. **b** and **c** Time-resolved photoreflectance (PR) spectra in the N- and I-phases, measured at 100 and 77 K, respectively. Δt denotes the interval between the excitation pulse and the white-light pulse for probing reflectance change. *Solid*, *dashed*, and *dotted lines* denote PR spectra observed well after ($\Delta t = 670$ and 820 ps), 300 ps after, and 100 ps before excitation, respectively. PR spectra were recorded as the relative difference ($\Delta R/R$) between the spectra with and without irradiation to a pulsed laser light of 80-fs pulse width. Excitation photon energy was 1.55 eV, and photon flux for excitation was 5×10^{13} and 4×10^{13} cm^2 for observations at 77 and 100 K, respectively [15]

(a) Time-resolved reflectance (absorbance) measurement

As a result of recent progress in laser technology, we can easily utilize pulsed laser light with 100 femto(10^{-15}) s width for time resolved spectroscopy. As well known, the intensity of fs laser light is very high (>G (10^9) W) though the total pulse energy is relatively low and thus fs-laser light can easily induce nonlinear optical effect in various transparent matter. Utilizing one of nonlinear optical effects, so called as self-phase modulation (SPM) effect [14], we can obtain white light pulse with about 150 fs width which is profitable for monitor light probing the change in optical properties of target materials. In Fig. 8.21, the experimental apparatus to study the time-resolved reflection spectral change at various temperatures with 150 fs time resolution based on this pump-probe technique. The sample surface is excited by fs laser pulse of which photon-energy can be tuned by optical parametric technique. The change in reflectance at various photon-energy after fixed delay time (Δt) is probed by white pulsed light generated by SPM effect in sapphire crystal and analyzed by polychromator-CCD coupled system1. Δt value can be changed by the change in path-length of

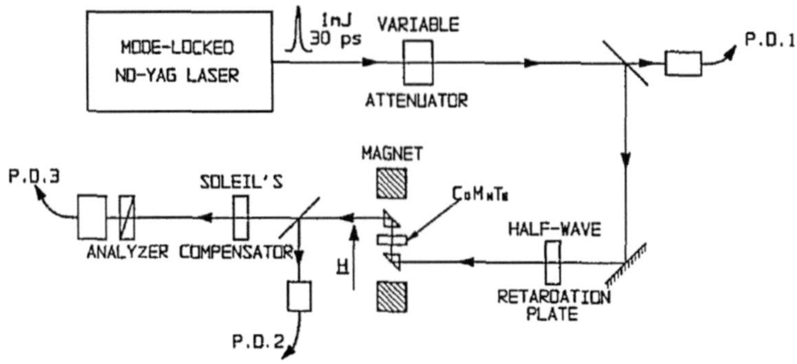

Fig. 8.23 Experimental setup for the measurement of magneto-optical properties which are nonlinearly depending on the light intensity [16]

optical delay-line. The 2nd coupled polychromator and CCD system is utilized for the calibration of white light spectrum and intensity of it. Figure 8.22 shows the example of time-resolved reflectance spectra observed for organic charge transfer crystal TTF-CA [15]. Of course, the position of optical delay line for $\Delta t = 0$ should be carefully determined. Because as a result of chirp effect in probe white pulse due to photon-energy dependence of refractivity in optical elements, the position for $\Delta t = 0$ will depend on the probe photon-energy. In some exaggerated case, such effect reaches more than 100 fs depending on the utilized optical parts in experimental set-up. For the calibration of $\Delta t = 0$ for probe light, the nonlinear optical effect such as sum-frequency light generation is utilize [14].

(b) Time-resolved measurement method for MO property
In previous section, we have mainly discussed about the method for probing the dynamical change in reflectivity and absorbance. In other words, we mainly treated about the change in the intensity of probe light due to photo-excited species. Of course, another important nature of light is the polarization. The change in polarization of probe light is strongly correlating with the dynamical behaviors of structural and/or magnetic properties induced by photo-excitation. Here, we focus on the measurement of dynamics of MO property. There are two methods for the study of pico-second time-resolved change in polarization nature of photo-excited materials. The first one, which is most convenient way, is the nonlinear MO property measurement using one light source variously changing the light intensit [16]. Figure 8.23 shows the typical experimental set up for this method. The laser intensity is adjusted by the use of a variable attenuator and the linear incident polarization vector is rotated by half-wave retardation plate. The attenuation of irradiated laser pulse is monitored by photodiodes 1 and 2. The state of polarization and the phase of the transmitted light can be estimated by rotating the analyzer (polarization rotation) and varying the thickness of the Soleil compensator (phase change), respectively.

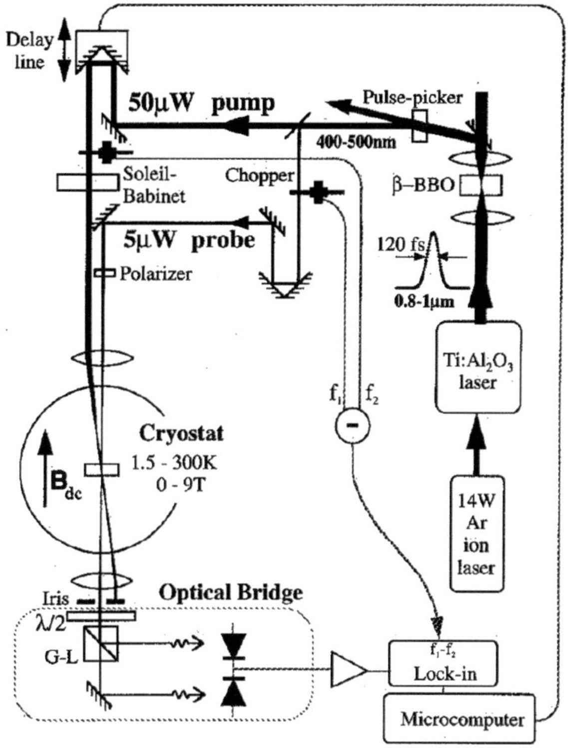

Fig. 8.24 Schematic diagram of the time-resolved Faraday rotation experiment with 120 femto-second resolution. Pulse-picker is utilized to reduce the repetition rate of the laser [17]

The other method for measurement is the use of pico- and femto-second pulsed laser in a pump-probe configuration. Following excitation by a polarized pump pulse, the change in Kerr rotation angle ϕ and/or Kerr ellipticity η is measured as a function of delay time. Figure 8.24 shows the typical example for the time-resolved Faraday rotation measurement [17]. Laser pulse with 120 fs width is split into pump and probe light. The photo-induced Faraday rotation can be probed as a result of the change in the ratio between the light intensities for $X(I_x)$ and $Y(I_y)$ polarization directions. Before photo-excitation or at negative delay time (i.e. the probe pulse reaches sample just before pump pulse), the I_x and I_y are adjusted to the same value (balanced state). If the Faraday rotation angle is modulated by pump pulse at positive delay time (i.e. the probe pulse reaches sample just after pump pulse), the discrepancy between I_x and I_y is induced. Therefore, we can sensitively probe the change in Faraday rotation angle by detecting the differential signal utilizing lock-in technique referenced to the difference of the pump and probe modulation frequencies. Typical example for the measurement of the photo-induced Faraday rotation is plotted in Fig. 8.25. As shown in this figure, Faraday rotation less than 1 mdeg. can be sensitively probed by this method with the time-resolution of 120 fs.

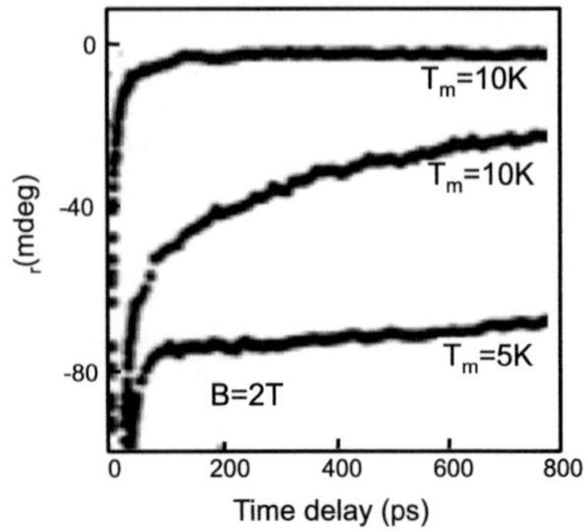

Fig. 8.25 Temperature dependencies of the pump-induced Faraday rotation taken at constant magnetic field observed for thin films with magnetic heterostructure [17]

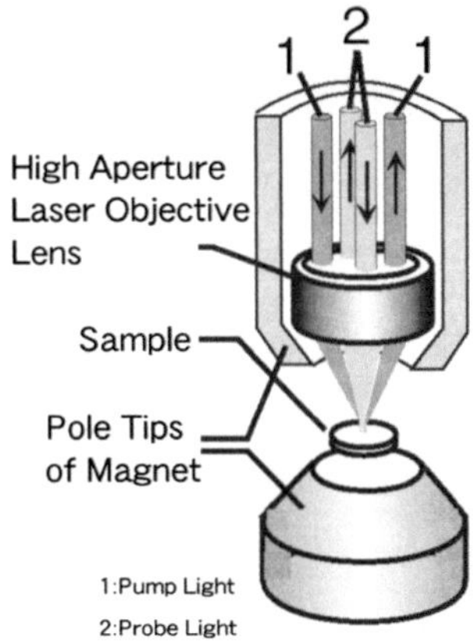

Fig. 8.26 Schematic view graph of the experimental configuration for the measurement of the microscopic photo-induced magneto-optical effect. Pump and probe beams are denoted as 1 and 2 respectively [18]

Recently, this pump-probe technique for MO property measurement is expanded into microscopic measurement [18]. Figure 8.26 shows the experimental set-up for the microscopic MO measurement. Inside the hollow pole tip of an electromagnet, a high aperture objective lens is mounted and spatial resolution of about 5 μm is achieved.

References

1. S. Huard, Stress-induced or stress-modified anisotropy, Chapter IV, in *Polarization of Light* (Wiley, New York, 1997)
2. M.C. Gupta, Photonic devices and optics, Part B, in *Handbook of Photonics* (CRC Press, New York, 1997)
3. R.W. Ditchburn, Anisotropic media, in *Light* (Dover Publications Inc., New York, 1991)
4. T. Komori, T. Ishikawa, T. Kuroda, J. Yoshino, F. Minami, S. Koshihara, Phys. Rev. B **67**, 115203 (2003)
5. S. Sugano, N. Kojima, *Magneto-Optics*, Springer Series in Solid-State Physics 128 (Springer, Berlin, 2000)

6. J.A. Gaj, Magneto-optical properties of large-gap diluted magnetic semiconductors, Chapter 7, in *Diluted Magnetic Semiconductors-Semiconductors and Semimetals*, ed. by J.K. Furdyna, J. Kossut, vol. 25 (Academic Press, San Diego, 1988)
7. S. Takeyama, Spin-Induced optical phenomena in diluted magnetic semiconductors, in *Optical Properties of Low-dimensional Materials*, ed. by T. Ogawa, Y. Kanemitsu, vol. 2 (World Scientific, Singapore, 1998)
8. J. Szczytko, W. Bardyszewski, A. Twardowski, Phys. Rev. B **64**, 075306 (2001)
9. K. Sato, Jpn. J. Appl. Phys. **20**, 2403 (1981)
10. Catalog of Streak Camera, Hamamatsu Photonics (2003)
11. S. Tomimoto, S. Saito, T. Suemoto, J. Takeda, S. Kurita, Phys. Rev. B **66**, 155112 (2002)
12. S. Takeyama, Y. Takagi, T. Karasawa, V.F. Aguekiand, J. Cryst. Growth **184/185**, 917 (1998)
13. H. Okamoto, K. Egawa, M. Takahashi, Y. Oka, J. Lumin. **76/77**, 211 (1998)
14. Y.R. Shen, *The Principles of Nonlinear Optics* (Wiley, New York, 1984)
15. S. Koshihara, Y. Takahashi, H. Sakai, Y. Tokura, T. Luty, J. Phys. Chem. B **103**, 2592 (1999)
16. J. Frey, R. Frey, C. Flytzanis, R. Triboulet, J. Opt. Soc. Am. B **9**, 132 (1992)
17. S.A. Crooker, D.D. Awschalom, N. Samarth, IEEE J. Sel. Top. Quantum Electron. **1**, 1082 (1995)
18. B. Koopmans, M. van Kampen, J.T. Kohlhepp, W.J.M. de Jonge, Phys. Rev. Lett. **85**, 844 (2000)

Index

Y. Aoyagi and K. Kajikawa (eds.), *Optical Properties of Advanced Materials*,
Springer Series in Materials Science 168, DOI: 10.1007/978-3-642-33527-3,

MIX
Papier aus verantwortungsvollen Quellen
Paper from responsible sources
FSC® C105338

If you have any concerns about our products,
you can contact us on
ProductSafety@springernature.com

In case Publisher is established outside the EU,
the EU authorized representative is:
Springer Nature Customer Service Center GmbH
Europaplatz 3, 69115 Heidelberg, Germany

Printed by Libri Plureos GmbH
in Hamburg, Germany